How to Open a Financially Successful Bakery

With Companion CD

– Revised 2nd Edition –

Sharon L. Fullen and Douglas R. Brown

Revised and Updated by **Zachery Humphrey**

HOW TO OPEN A FINANCIALLY SUCCESSFUL BAKERY WITH COMPANION CD REVISED 2ND EDITION

Copyright © 2015 Atlantic Publishing Group, Inc.
1405 SW 6th Avenue • Ocala, Florida 34471 • Phone 800-814-1132 • Fax 352-622-1875
Web site: www.atlantic-pub.com • E-mail: sales@atlantic-pub.com
SAN Number: 268-1250

Library of Congress Cataloging-in-Publication Data

Fullen, Sharon L.
  How to open a financially successful bakery / Sharon L. Fullen and Douglas R. Brown. -- Revised 2nd edition.
     pages cm
  Includes bibliographical references and index.
  ISBN 978-1-60138-941-1 (alk. paper) -- ISBN 1-60138-941-8 (alk. paper) 1. Bakeries--United States. 2. New business enterprises--United States--Management. 3. Small business--United States--Management. I. Brown, Douglas Robert, 1960- II. Title.
  HD9057.U59F85 2015
  664'.7520681--dc23
                                   2014034828

EDITOR: Melissa Figueroa • mfigueroa@atlantic-pub.com
INTERIOR LAYOUT: Antoinette D'Amore • addesign@videotron.ca
COVER DESIGN: Meg Buchner • meg@megbuchner.com
JACKET DESIGN: Jackie Miller • sullmill@charter.net

Printed on Recycled Paper

Printed in the United States

Reduce. Reuse.
RECYCLE.

A decade ago, Atlantic Publishing signed the Green Press Initiative. These guidelines promote environmentally friendly practices, such as using recycled stock and vegetable-based inks, avoiding waste, choosing energy-efficient resources, and promoting a no-pulping policy. We now use 100-percent recycled stock on all our books. The results: in one year, switching to post-consumer recycled stock saved 24 mature trees, 5,000 gallons of water, the equivalent of the total energy used for one home in a year, and the equivalent of the greenhouse gases from one car driven for a year.

Over the years, we have adopted a number of dogs from rescues and shelters. First there was Bear and after he passed, Ginger and Scout. Now, we have Kira, another rescue. They have brought immense joy and love not just into our lives, but into the lives of all who met them.

We want you to know a portion of the profits of this book will be donated in Bear, Ginger and Scout's memory to local animal shelters, parks, conservation organizations, and other individuals and nonprofit organizations in need of assistance.

– Douglas & Sherri Brown,
President & Vice-President of Atlantic Publishing

Disclaimer

• •

The material in this book is provided for informational purposes and as a general guide to starting a bakery only. Basic definitions of permits, licenses, and rules are provided according to the current status of the laws and information at the time of printing; be sure to check for a change or update in health and food-handling laws. This book should not substitute professional and legal counsel for the development of your bakery business.

Table of Contents

Chapter 1

Starting Your Own Bakery

• •

Operating a bakery is hard work, and the failure rate unfortunately is quite significant. However, realizing your dream can be an exceptionally rewarding endeavor – personally, as well as professionally. The ingredients for success go well beyond an exquisite cheesecake or hearty oat bread. They are a complex blend of passion, vision, risk taking, and sharpness.

Solidifying Your Vision

Your dream of owning a business sparks your quest, but creating a solid vision of how you want to accomplish your dream is truly the foundation for success. Creating a vision will help advance you towards writing your business plan, selling your concept to lenders and potential investors, and communicating your desires and needs to architects, contractors, designers and suppliers.

Below are some ways to help you solidify your vision. This fundamental understanding will help you correctly make the tough decisions when faced with compromises, budgetary problems and unforeseen obstacles.

Ways to Explore Your Passion

- **Close your eyes.** Envision your future business. Can you see your dream bakery? Are the glass displays filled with artfully arranged pastries? Do regulars gather every weekday for a Danish pastry and cup of coffee? Are parents bringing their kids in for a few breakfast bagels? Can you smell enticing aromas drifting in from the kitchen? Do your taste buds tingle when you think of your own personal specialty? These images all represent your passion.

- **Take a few hours of uninterrupted time to think** about your personal and financial reasons for committing your energies and "nest egg," tying yourself to a long-term loan or even taking on a business partner. It takes time to become profitable – are you financially and emotionally prepared for this kind of investment?

- **Create a list of the positives and negatives.** Every venture has risks but the positives should always outweigh them. If your entrepreneurial spirit isn't damped by the potential risks, then you are ready to take the next step – giving your vision a voice.

- **Determine what talents you can personally bring to the process.** Do you love to bake or are you more interested in simply selling baked goods? Do you want to write a check and let the professionals handle the details or do you want to be consulted on every detail from the front door to the walk-in freezer?

- **Write a one-minute "elevator pitch."** If you found yourself in an elevator with a wealthy investor, how would you describe your vision (and secure the cash) in the time it takes to travel up 20 floors? Show your passion while simultaneously emphasizing the tangible benefits!

Launching Your Business

There are three ways to launch into business for yourself: 1) start from scratch, 2) invest in a franchise or 3) buy an existing bakery. In Chapter 2 ("Buying a Bakery") and Chapter 3 ("Buying a Franchise"), you'll learn the advantages and disadvantages of each. In Chapter 4, you'll also learn about properly preparing a business plan.

All three methods (scratch, franchise, existing) can be the basis for a successful bakery business. The differences between each are as much emotional and psychological as they are financial. By weighing the pros and cons, factoring in your personality and business expectations and then balancing it with your potential customer's needs and desires – you can decide the best method for you.

Jump Starting a Business

Purchasing an existing business can be the fastest way to get your doors open; however, it may not be the best choice for success. Remember that if the location is poor, the prior business either had a bad reputation or the equipment is overpriced – which may hinder your potential.

Availability is another factor when purchasing an existing business. The right business for you has to be on the market *now* or you have to make an offer that is so substantial that they cannot refuse.

Quality franchise organizations have market restrictions, assigned territories and several other ways to keep their franchisees from competing or flooding the market. This may mean that your desired franchise is unavailable in your area. Popular franchises often require higher investments and a great personal net worth.

Starting from Scratch

There's nothing like a cake made from scratch – in fact that may be why you want to open your own bakery. Starting your own bakery from the ground up gives you the opportunity to select every cabinet, paint color, mixer, and employee – to create a business that closely reflects your individuality. However, like overbeaten batter, it can actually turn out to be too much of a good thing and go far beyond your scope of experience or budget.

If you have taken time to do some personal reflection, compared your resources to your needs, and set realistic expectations – you'll be better prepared to build, buy or lease your bakery facility, launch a new business and create a foundation for success and profits. The worksheet below is a helpful assessment tool to determine which way of starting a business best suits you.

GOING INTO BUSINESS			
(This form can be used to tally pluses and minuses or to make comments.)			
Factors	From Scratch	Buy Existing Business	Franchise
Time			
Availability			

Launch time (from Planning to Opening)			
Financial			
Cost			
Available Financing			
Investors			
Personal Worth Requirements			
Total Indebtedness			
Breakeven Point			
Royalties & Fees			
Purchasing Restrictions			
Current Profitability			
Intangibles			
Goodwill			
Historical Recognition			
Known vs. Unknown (obstacles to success or existing profitability)			
Reputation			
Convenience			
Exclusivity			
Assets			
Location			
Facility			
Equipment			
Existing Staff			
Customer Base			

Owner			
Independence			
Business Experience			
Food Service Experience			
Bakery Experience			
Management Experience			
Owner Expectations			
Outside Expectations			
Training			
Support			
Marketshare			
Marketing Support			
Product Mix			
Competition			
Customer Needs			
Other (your personal list)			

Don't Know the Answers?

If you don't know the pros and cons of the three ways to launch a business, you will find it helpful to immediately start researching your market, the competition and existing business opportunities. You'll have to do this BEFORE you prepare a formal business plan so that the information you gather will help you prepare.

Chapter 2

The Basics of Buying a Bakery

B uying an existing bakery has its advantages and disadvantages. The prior bakery's track record of success can be a good reference point for your potential in the location. As with most food service businesses, location plays a vital role in your success. Before purchasing an existing bakery, carefully review all financial records, have appraisals done and, most importantly, consult a lawyer.

This chapter addresses many of the issues that are important to you – the buyer. When it's your turn to become the seller, you'll benefit from understanding everyone's role in buying and selling your bakery business. Both sides of the transaction will be discussed here, and you can learn more about selling your bakery in Chapter 24.

Real Estate and Its Value

Real estate is land and any permanent improvements made on the land, such as utility connections, parking lots, buildings, etc. The real estate property of a bakery is often its most valuable asset. However, many bakeries are bought and sold without the real prop-

erty being part of the operation because the business is operated in a leased building. Generally, there are three procedures for determining the value of a bakery: market approach, cost approach, and income approach.

Market Approach – the value of a property is determined by comparing it to like pieces of property in similar areas. The market approach is generally not used to estimate the value of a bakery's real estate. However, if you are selling land only or the land portion of a bakery, the market approach can be an accurate determiner of value, since there are considerably fewer adjustments to make.

Cost Approach – the property is valued on what it would cost to replace it completely. The costs are based on current purchase prices for new equipment and assets. In the case of equipment that is no longer made, add the price of what a new piece of equipment that provides the same utility would cost. Include all taxes, freight and installation in your quotes, and factor in depreciation. The cost approach is not widely used to estimate the value of a bakery's real estate, but is used mostly by insurance companies while processing a claim.

Income Approach – the value is based on future income to be derived from the property. The real estate value, then, is the present value of the estimated future net income, plus the present value of the estimated profit to be earned when the property is sold. This is the preferred approach when determining an accurate sales price for an income-producing property.

The Value of Other Assets

The business itself consists of everything the owner wishes to sell. Usually this means furniture, fixtures, equipment, leasehold improvements, etc. It may also include tax credits, favorable operating expenses, customer lists and name recognition. The price for a food-service business is usually 40–70 percent of the operation's 12-month food-and-beverage sales volume. The seller usually will set the sale price at the high end of this percentage, and the prospective buyer will set it at the low end.

Setting a sales price is not, of course, a straightforward process, and there are many other factors that need to be considered. Here are a few:

Profitability

This has the most influence on sales price and salability of a bakery. The most common way to determine profitability is to examine the net operating income figure and compare this to industry standards and the regional standard for that type of operation.

Leasehold Terms and Conditions

The term remaining on the property lease and monthly payment will affect the sales price greatly. Avoid anything less than a five-year lease unless it's very profitable or priced very low. Most buyers also need a reasonable monthly payment and a reasonable common area maintenance payment. This should not exceed 6 to 8 percent of the monthly food-and-beverage sale, if at all possible.

Track Record

Businesses need to show acceptable track records to entice buyers. This usually means the business must be at least a year old. The track record will then be used to project the business's future prospects. If a business depends on the work of highly skilled employees such as a well-known pastry chef, this too can affect the price because it makes the business more difficult to expand and more expensive to operate.

Other Income

Most bakeries don't earn significant other income – often less than 2 percent. However, there are rebates, interest on bank deposits, vending machines, salvage from aluminum, grease and cardboard, etc. You may also consider gaining other income by selling seats at potential meetings or conferences with other small business owners in order to strengthen your business when it is in a strictly local phase.

Below-Market Financing

When a bakery is sold, usually the buyer puts up a small down payment and the seller then carries back the remainder of the sales price at favorable terms. Seller financing is almost always below market, and the buyer avoids the fees associated with bank loans.

Franchise Affiliation

If a bakery is part of a large franchise, the sale price will increase significantly. This is truer of the larger national franchises than the regional ones.

Number of Buyers and Sellers

A seller should plan to market his business when there are as many potential buyers as possible. This means early spring and summer—especially if it's a tourist business—or after legislation limiting construction has passed, taxes are lowered, etc.

Contingent Liabilities

Contingent liabilities reduce a bakery's net income. These may be coupons issued by the previous owner pension plans that eat into your net profit margin. If a buyer cannot eliminate these expenses, they most likely represent a negative value that should be factored into the offering price.

Grandfather clauses

New owners are expected to meet fire, health and safety codes that the previous owner may have been able to avoid because of being "grandfathered in" when the regulations were passed. Grandfather clauses usually expire when a business changes hands. If this is the case, the seller or buyer may need to bring the building up to code. If the buyer is responsible for this, she will usually ask that the expense be deducted from the sale price. If the cost is very high, this could affect the salability of the business altogether.

Goodwill

The IRS determines goodwill as the amount of money paid for a bakery in excess of the current book value of the physical assets. It essentially determines the value of your business based on expected continued customer patronage specifically due to its individual name, reputation, or any other factor. Most investors however look at excess earnings as attributable to positive goodwill and deficient earnings to negative goodwill. A buyer may be willing to pay for goodwill, but a seller should expect the buyer to downplay its value in order to lower the sale price as much as possible.

Terms, Conditions and Price

In most cases, sellers will determine a likely sales price, terms, and conditions then pad those somewhat to create room for negotiation and compromise. Employing a skilled business broker can help determine your price offer.

Terms

The terms of sale are the procedures used by the buyer to pay the seller. Seller financing is a frequent arrangement where you will make a minimal down payment and pay the remainder over a 3- to 5-year period. All-cash offers are rare so the seller probably won't be seeking that. It is the seller's interest to seek a large down payment. Your financial advisor can assist you in determining how much down payment you should be willing to make. Sellers are also more likely to grant favorable terms to a buyer making a substantial down payment because the financial risk is lessened.

Conditions

There are several conditions the seller and buyer will attach to most sales contracts. Sometimes they are separate agreements, but most of the time they are part of the sales contract. The following are conditions that are the largest concern to the seller:

Conclusion of sale – the amount of time/how long it will take to close the sale

Buyer access – your access to the facility and staff to facilitate a smooth transition

Guarantees – sellers usually have to guarantee the condition of assets. Sometimes sellers have to guarantee that buyers can assume some of the bakery's current contracts. Imprecise language should also be avoided here. If the seller is making guarantees, then buying the relevant insurance to back up these claims is prudent

Indemnification – the conditions and penalty should anyone back out of the deal

Escrow agent – agents hired to supervise the sales transaction paperwork. It is wise to use an independent third-party agent

Legal Requirements – agreements to comply with all pertinent laws and statutes

Buyer's credit history – providing personal and business financial records and permission to run a credit report

Security for seller financing – a clause requiring the sellers approval before the new owner can obtain additional financing the promissory note will contain a default provision that the lender can foreclose if loan payments are not met, in addition to other specific provisions pertinent to the business

Assumable loans and leases – detailing assumable contracts, loans and leases

Life and disability insurance – insurance naming the seller as beneficiary should the buyer be unable to pay the loan due to death or disability

Collection of receivables – a fee paid to the seller if they collect receivables incurred by the seller but payable to the new owner

Inventory sale – a physical inventory of all food, beverages and supplies taken at the end of escrow; Use an independent service and a separate bill of sale prepared for the agreed-upon price of this merchandise.

Non-compete clause – an agreement that the seller will not open up a competing business near by; usually a time period is set and "competing business" is defined

Repurchase agreement – agreements that grant the seller an option to buy the bakery back within a certain time period

Employment contract – details of a seller becoming an employee of the new owner

Consulting contract – an alternative to the employment contract; This may be a more acceptable employment contract that gives the new owner a tax-deductible expense but doesn't burden the former owner, either.

Conditions not met – detail if the seller can back out do to the buyer not meeting the sales conditions

Initial Investment

Buyers must estimate, as accurately as possible, the total initial investment needed to get their businesses up and running the way they envision them. Many bakeries that could have been successful failed because they were undercapitalized. For exactly this reason, one of the very

appealing aspects of purchasing an existing bakery is that many start-up costs are avoided. There are, however, a number of start-up costs even with transfer of owner-ship. Here are a number to be aware of:

Investigation costs – must be willing to spend time and money to thoroughly examine the opportunities that are available; Many investors falsely believe that once initial development work is complete, the start-up costs are eliminated. These costs still exist and wise investors calculate them in their analyses.

Down payment – a standard down payment is usually around a quarter of the sales price. The down payment can affect the sales price, and in many cases sellers will accept a lower sales price with a larger down payment and vice versa.

Transaction costs – prorated insurance, payroll, property taxes, vacation pay, license renewal fees, advertising costs, etc. on the close-of-escrow date.

Working capital – available cash to ensure sufficient supplies are on hand to run the bakery.

Deposits – cash deposits required of the new owner for utility, telephone, sales tax, payroll-tax and lease deposits.

Licenses and permits – all required operating licenses and permits for retail, health and occupational

Legal fees – fees for legal advice, buyer negotiate and contract review.

Renovations – costs required to renovate or rectify building code violations.

Equipment and utensils – costs to purchase new or replacement equipment, supplies and serving ware (Don't forget to include maintenance agreements!)

Advertising – costs to promote an opening or reopening, rebuild signage and offer promotional discounts and incentives.

Fictitious name registration – also known as "doing business as" or DBA; If the name of a bakery doesn't use your own name, the name usually must be registered at the local courthouse or County Recorder's Office.

Loan fees – loan fees from the lending parties.

Equity fees – attorney, document preparation and registration fees for selling common stock

Insurance – lender required life and disability insurance with the lender named as sole   beneficiary (Adequate real property insurance may also be required).

Franchise fees – franchise transfer-of-ownership fee; This fee pays the franchiser for the costs of evaluating the new owner for the franchise. It is paid in up front and in cash before the new franchise can begin operations.

Distributorship fees – exclusive distributorship licenses or discontinuing a current license agreement may incur costs similar to franchise fees

Pre-opening labor – labor required during pre-opening and transition period

Accounting fees – fees for assistance in the evaluation of a bakery purchase

Other consulting fees – fees for specialty services such as bakery consultants, labor-relations specialists and computer consultants

Other prepaid expenses – any prepayment required by a creditor

Sales taxes – property subject to a transfer tax & non-food supplies are often subject to sales tax

Locksmith – cost to change all the locks on a bakery after the sale is concluded

Security – transfer or set-up fee for security service or systems

Contingency – a contingency fund large enough for at least the first six months' operating expenses; Among other things, it is often necessary to over-hire and over-schedule employees before an effective sales distribution pattern emerges, so operators incur incredibly high expenses during the first six months of operation. Not having an ample contingency fund is the primary reason many businesses fail within the first few months.

Strategies for Buying a Bakery

You must determine the type of bakery that is right for you. Consider the bakery's investment yield, taxes and the effect that the business will have on your personal life. Basically, buyers should be looking for bakeries that will meet their numbers, and workers that they are going to be happy working in the wee hours and maybe 12–14 hours a day.

Reviewing the Books

As a potential buyer, you thoroughly analyze the finances of the bakery. Carefully study its current profitability and use this information to determine its potential for generating revenue.

It's a good idea for you to hire a good accountant to assist in determining whether the deal meets your investment requirements. If the buyer qualifies, Small Business Development Centers (SBDC) throughout the country offer free consulting services to businesses with fewer than 500 employees.

You should also complete at least a rough market and competition survey before performing your financial evaluation. This will ensure your familiarity with the bakery's location and will help estimate future revenues and expenses. If you are unfamiliar with the area, an independent consulting service can provide a useful survey.

A seller expects a written offer with price, terms, conditions and an earnest money deposit before he or she will allow a potential buyer to review confidential financial information. It should be agreed that you could withdraw if he or she's unhappy with the financial records. This is because the earnest deposit is at risk unless the right to retract the offer is in place.

Cost of food, beverages and supplies is a bakery's biggest expense. The potential buyer should take a random sample of canceled invoices and check their consistency with the cost of goods sold and direct operating-supplies expenses listed on the current income statement. If these numbers match, you have a good idea of what product and supply expenses the bakery will incur if no organizational and operational changes are made. The canceled invoices are also a good test of the current owner's purchasing skills. If invoices show higher prices than those of competing suppliers, you can expect to decrease those expenses.

Most lenders require a cash budget to be prepared by potential buyers. This will point out the operation's daily cash requirements and the times of year when short-term money must be borrowed to cover brief shortages.

Potential buyers also should analyze balance sheets and income statements carefully. Balance sheets can reveal anxiety level of a seller and indicate the current management's ability. If this ability is in question, this could predict greater earnings under sound management. Income statements should be used to determine whether the bakery could have satisfied salary demands and provided a return on the initial investment, had it been under your management for the previous 12 months. As a buyer, you are *basing* your offer on current income as well as purchasing the operation's *future* revenue-making ability.

Your Objectives

As a buyer, you must prioritize your objectives and consider the trade-offs that may be made to attain them. Buyers generally want to accomplish the following objectives:

- **Best possible sales price**
- **Reasonable down payment**
- **Reasonable initial investment**
- **Maximum future profits**
- **Reduced possibility of failure**
- **Enhancement of borrowing power**
- **Minimized tax liabilities**

Financing

Your qualification for financing and all necessary permits and licenses are the two most common obstacles to purchasing a business. There is little you as a buyer can do if you don't qualify for permits, unless you need to fix only simple code violations to qualify. In the financing realm, however, the seller and buyer have more control and can adjust the final sales contract to suit the buyer's needs. Buyers should beware that there is no bigger threat to their success than inadequate or inappropriate financing. Excessive debt burden is one of the most consistent reasons why bakeries go under.

Equity Funds

Equity is the capital that is at risk. It's basically the money that you need to commit to in order to start your business. Owners invest their equity funds without any guarantee on a return. There are several types of equity financing techniques. The following are the most common:

Personal equity – funded entirely with personal equity or with a combination of personal equity and lease and debt financing.

Partnerships – partners investing together; these may be partners who participate in the business or "silent" inactive partners interested in a passive investment.

Corporations - raising capital through the sale of stock to private investors or the public

Venture capital – using professional investors or investment companies; Venture capitalists are interested in long-term financial gain and are less interested in the net operating profits of a new establishment. Buyers intending to purchase chain bakeries might be able to obtain venture capital because of the high earning potential.

Borrowing Money

It is often said that small-business owners have a difficult time borrowing money. This is not necessarily true. Banks earn money every year just by lending money. However, the financial inexperience of small business owners often prompts banks to deny loan requests. Requesting a loan when you are not properly prepared sends a warning signal to your lender. That message reads: High Risk!

To be successful in obtaining a loan, you must be prepared and organized. You must know exactly how much money you need, why you need it and how you will pay it back. You must be able to convince your lender that you are a good credit risk. Your loan request will be supported by your formal written business plan. Chapter 4 discusses writing this important document.

How Your Loan Request Will Be Reviewed

When reviewing a loan request, the lender is primarily concerned about repayment. To determine your ability to repay, many loan officers will order a copy of your business credit report from a credit-reporting agency. Using the credit report and the information you have provided, the lending officer will consider the following issues:

- Have you invested savings or personal equity in your business totaling at least 25 percent to 50 percent of the loan you are requesting? (Remember, a lender or investor will not finance 100 percent of your business.)

- Do you have a sound record of credit-worthiness as indicated by your credit report, work history and letters of recommendation? This is very important.

- Do you have sufficient experience and training to operate a successful business?

- Have you prepared a loan proposal and business plan that demonstrate your understanding of, and commitment to the success of, the business?

- Does the business have sufficient cash flow to make the monthly payments?

SBA Financial Programs

The Small Business Administration (SBA) offers a variety of financing options for small businesses. Whether you are looking for a long-term loan for machinery and equipment, a general working capital loan, a revolving line of credit or a micro-loan, the SBA has a financing program to fit your needs. These programs are discussed in detail on SBA's website at **www.sba.gov**.

U.S. Small Business Administration

The 7(a) Loan Guaranty Program is the SBA's primary loan program. The SBA reduces risk to lenders by guaranteeing major portions of loans made to small businesses. This enables the lenders to provide financing to small businesses when funding is otherwise unavailable at reasonable terms. The eligibility requirements and credit criteria of the program are very broad in order to accommodate a wide range of financial needs.

The SBA also offers a few other loan programs including a Microloan Program (small, short-term loans), CDC/504 Program (financing for major fixed assets such as equipment or real estate), and Disaster Loans Program (low-interest loans for repairing/replacing various things damaged in an unfortunate, unforeseen circumstance). Each loan program has different requirements, amounts, and components that you can check at **www.sba.gov/loanprograms**.

In order to determine whether you qualify or whether an SBA business loan best suits your financing needs, contact your banker, one of the active SBA-guaranteed lenders or an SBA loan officer.

> **Friends and Relatives** – private sources such as friends and family willing to grant interest-free or low-interest rates

> **Banks and Credit Unions** – the most common sources of funding, banks and credit unions. Loans are based on solid business proposals and your written business plan

Closing the Sale

Once you and the seller have agreed on the terms, you will sign a binding sales contract and transfer ownership. There are often lawyers, brokers, accountants, lenders, escrow agents, government officials, trade unions, family members and other people involved in this transaction. It usually takes 30–60 days to finish the ownership transfer. After everything has been processed, congratulate yourself—you just completed the first step of making your dreams come true!

Chapter 3

How to Invest in a Franchise

. . . .

$\mathcal{M}$any bakery owners have started their career by investing in a franchise. According to the U.S. Department of Commerce, buying a franchise is the average person's most viable avenue to owning a business. You may want to consider such an investment. Franchising can minimize your risk. It will enable you to start your business under a name and trademark that has already gained public acceptance. You will have access to training and management assistance from experienced people in the bakery industry. Sometimes, you can obtain financial assistance; this allows you to start your business with less cash than you would ordinarily need.

On the other hand, you must make some sacrifices when entering a franchised operation. You lose a certain amount of control of the business. You will no longer truly be your own boss in some situations. You can be adversely affected by the actions of the franchisor (negative publicity, bankruptcy). And, of course, you must pay a fee or share profits with the franchisor. This chapter will present some of the **advantages** and **disadvantages** of franchising and how to evaluate a franchise opportunity.

Definition of Franchising

Essentially, franchising is a plan of distribution under which an individually owned business is operated as though it were a part of a large chain. Products are standardized, such as trademarks, symbols, design elements and equipment. A supplier (the franchisor) gives the individual dealer (the franchisee) the right to sell, distribute and market the franchisor's product by using the franchisor's name, reputation and selling techniques.

The franchise agreement (or contract) usually grants the franchisee the exclusive right to sell, or otherwise represent the franchisor, in a specified geographical. In return for this exclusive right, the franchisee agrees to pay a sum of money (a franchise fee) or a percentage of gross sales and/or to buy equipment or supplies from the franchisor—often these options are variously combined.

Advantages of Franchising

As a franchisee, you have the luxury of starting a business with:

1. **Limited experience.** You are taking advantage of the franchisor's experience, experience that you probably would have gained the hard way—through trial and error.

2. **Reduced capital outlay and a strengthened financial/credit standing.** Some franchisors provide financial assistance to establish your bakery with less than the usual cash investment. With the name of a well-known, successful franchisor behind you, your standing with financial institutions can also be strengthened.

   a. For example, the franchisor may accept a down payment with your note for the balance of the needed capital. Or the franchisor may allow you to delay in making payments on royalties or other fees

3. **Well-developed image and consumer recognition of proven products and services.** The goods and services of the franchisor are typically proven and widely known. Therefore, your business has "instant" pulling power. To develop such pulling power on your own might take years of promotion and considerable investment.

4. **Competently designed facilities, layout, displays and fixtures.** The franchising company has designed standardized facilities, layouts, displays and fixtures based upon their experience with many dealers.

5. **Chain buying power.** You may receive savings through chain-style purchasing of products, equipment, supplies, advertising materials and other business needs.

6. **Business training and support.** The franchisee's training program and ongoing business development support system teaches you proven business methods and gives you a valuable safety net. You can normally expect to be trained in the mechanics of the bakery business and guided in its day-to-day operation until you are proficient at the job. Moreover, management-consulting services are provided by the franchisor on a continuing basis. This often includes help with recordkeeping as well as other accounting assistance. Remember your success (and potential failure) directly reflects on their corporate image.

7. **National or regional promotion and publicity.** The national or regional promotion of the franchisor will help your business. In addition, you will receive help and guidance with local advertising. The franchisor's program of research and development will assist you in keeping up with competition and changing times.

These factors can increase your income and lower your risk of failure.

Disadvantages of Franchising

1. **Submission to imposed standardized operations.** You cannot make all the rules. Contrary to the "be your own boss" lures in franchise advertisements, you may not be your only boss. In addition, you must subjugate your personal identity to the name of the franchisor. Obviously, if you would like your operation to be known by your own name, a franchise is not for you. The franchisor exerts fundamental control and obligates you to:

   a. Conform to standardized procedure.

   b. Handle specific products or services that may not be particularly profitable in your marketing area.

c. Follow other policies that may benefit others in the chain but not you. This means that you forfeit the freedom to make many decisions – to be your own boss.

2. **Sharing of profits with the franchisor.** The franchisor nearly always charges a royalty (a percentage of gross sales). Frequently, this royalty fee *must be paid whether the franchisee makes a profit or not.* Sometimes such fees are exorbitantly out of proportion to the profit.

3. **Required purchases.** Merchandise, supplies or equipment that the franchisor requires you to buy from the corporation might be obtained elsewhere for less. A federal government study showed that in food franchising, many franchisees, that were required to buy a large proportion of supplies from their franchisors, were paying *higher prices* than they could obtain on their own. Additionally, you might pay more to the franchisor than other franchisees for the same services.

4. **Lack of freedom to meet local competition.** Under a franchise agreement, you may be restricted in establishing selling prices; in introducing additional products or services; or dropping unprofitable ones, even in the face of insidious local competition.

5. **Danger of contracts being slanted to the advantage of the franchisor.** Clauses in some contracts imposed by the franchisor provide for unreasonably high sales quotas, mandatory working hours, cancellation or termination of the franchise for minor infringements, and/or restrictions on the franchisee in transferring his franchise or recovering his investment. The territory assigned the franchisee may overlap with that of another franchisee or may be otherwise inequitable. In settling disputes of any kind, the bargaining power of the franchisor is usually greater.

6. **Time consumed in preparing reports required by the franchisor.** Franchisors require specific reports. The time and effort of preparing these may be inordinately burdensome. On the other hand, you should recognize that if these reports are helpful to the franchisor then he or she probably will help you to manage your business more effectively.

7. **Sharing the burden of the franchisor's faults.** While ordinarily the franchisor's chain will have developed goodwill among consumers, there may be instances in which ill will has been developed.

a.   For example, if a customer was served a stale roll or received poor service in one outlet he is apt to become disgruntled with the whole chain. As one outlet in the chain, you will suffer regardless of the excellence of your particular unit. Furthermore, the franchisor may fail. You must bear the brunt of the chain's mistakes as well as share the glory of its good performances.

Minority Participation in Franchising

A number of franchise systems have developed special programs for minority individuals who seek to go into business for themselves. One such program asks the minority individual for an only two-percent down payment. The franchisor matches this with 98 percent financing and up to a year of training.

Another program is a joint venture between a minority-owned business and an established franchising company. This joint venture is not a merger of the two companies. Rather, it is a plan whereby each company contributes an equal amount of dollars, but all responsibility for day-to-day operations is left with the minority-owned company.

Franchise Financing

There are a growing number of alternatives for individuals and investors who want to enter franchising or expand their current market position. More and more local and regional banks, along with national non-bank lenders, are offering franchise financing. Lending institutions have a greater appreciation for the importance of franchising in the marketplace, for its future growth and stability as a distribution method.

For example, the International Franchise Association (**www.franchise.org**) lists more than 30 bank and non-bank franchise lenders in its Franchise Opportunities Guide this year. The U.S. Small Business Administration (**www.sba.gov**), which last year backed more than 60,000 small-business loans totaling $14.75 billion, works with local and regional banks to offer its guaranteed loan program to start-up franchisees.

Evaluating a Franchise Opportunity

A franchise costs money. One can be purchased for as little as a few hundred dollars or as much as a quarter of a million dollars or more. Hence, it is vital that you investigate and evaluate carefully any franchise before you invest.

Beware of the "fast buck" artists. The popularity of franchising has attracted an unsavory group of operators who will take you if they can. Sometimes known as "front-money men," they usually offer nothing more than the sale of equipment and a catchy business name. Once they sell you the equipment, they do not care whether you succeed or fail. If you are promised tremendous profits in a short period, *be wary*.

The following checklist will aid you in selecting the right franchise. Check each question when the answer is "yes." Most, if not all, questions should be checked before you sign a franchise contract.

Franchise Research

To learn more about franchising and bakery (i.e. donuts, bagels, bread, cakes) franchise opportunities, check out the consulting, government and association websites listed below.

- Franchise Restaurant - **www.franchiserestaurant.com**
- Federal Trade Commission – research business opportunities **http://business.ftc.gov**
- International Franchise Association at **www.franchise.org**
- Wall Street Journal's Startup Journal at **www.startupjournal.com**
- Food Franchise.com - **www.foodfranchise.com**
- BISO – franchise advice at **www.bison1.com**

Conclusion

In conclusion, franchising creates distinct opportunities for a prospective small business owner. Without franchising, it is doubtful that thousands of small business investors could ever have started up. The system permits goods to be marketed by a small-business owner in a way that otherwise can done only with the vast sums of money and number of managerial people possessed by large corporations. As a new owner, you can draw from the experience and resources – promotional, managerial, and structural research – of a large, established parent company.

Unfortunately, not even the help of a good franchisor can guarantee success. You will still be primarily responsible for the success or failure of your venture. As in any other type of business, your return will be related directly to the amount and effectiveness of your investment in time and money.

Chapter 4

Planning for Success – Writing a Business Plan

no one plans to fail – but you can plan for success. Creating a formal written business plan is absolutely necessary if you are seeking outside financing. This plan is also your personal road map to success. Your plan will:

1. Help you clarify your ideas and dreams.

2. Share your dream with family and friends (especially if these are going to be financial backers or silent partners).

3. Focus your attention on building a solid business foundation.

4. Prepare you for potential obstacles and pitfalls.

5. Give you a reality check. Are you suited to the hours? Can you afford that fancy storefront? Does hiring employees scare you? Are you prepared for the impact it will have on your family life?

The Five Keys of Success

Before you can write anything down, you need to do some research within your community. You'll need to know whether you'll have all five of the keys to success.

1. **Need** – Does your community need a stand-alone bakery or another bakery? Is there a niche that is underserved?

2. **Customers** – Are there enough potential customers interested in your products to make you profitable? Will you have to spend heavily on advertising to reach them? Can they afford your products?

3. **Location** – Can your business be conveniently located? Can you afford the location? Are you near your ideal customer?

4. **Products** – Can you produce what your customers want at a price they will pay? Do you know what they want?

5. **Service** – Can you set yourself apart from competitors and make it worth the trip? Can you afford sufficient staffing?

Some of the answers to the above questions will require some research. We'll discuss each point in the order in which it would appear in your business plan.

What a Business Plan Includes

What goes into a business plan?

Your business plan can be divided into four distinct sections: 1) the description of the business, 2) the marketing plan, 3) the management plan, and 4) the financial management plan.

Addenda to the business plan should include the executive summary, supporting documents, and financial projections.

A Sample Business Plan Outline

Use the outline below as a guide to creating your first business plan. Throughout your first year and beyond, you should review and update your plan. Think about your short- and long-term goals; include what you've learned about your customers and competitors and how to keep your business growing.

Elements of a Business Plan

I.   Cover sheet

II. Statement of purpose

III. Table of contents

IV. The Business

    1.   Description of business

    2.   Marketing

    3.   Competition

    4.   Operating procedures

    5.   Personnel

    6.   Business insurance

    7.   Financial data

V.   Financial Data

    1.   Loan applications

    2.   Capital equipment and supply list

    3.   Balance sheet

    4.   Break-even analysis

    5.   Pro-forma income projections (profit & loss statements)

        i.   Three-year summary

        ii.   Detail by month, first year

        iii. Detail by quarters, second and third years

        iv.   Assumptions upon which projections were based

    6.   Pro-forma cash flow

        i.   Follow guidelines for number 5

VI. Supporting Documents

    1.   Tax returns of principals for last three years

    2.   Personal financial statement (all banks have these forms)

    3.   In the case of a franchised business, a copy of franchise contract and all supporting documents provided by the franchisor

    4.   Copy of proposed lease or purchase agreement for building space

    5.   Copy of licenses and other legal documents

    6.   Copy of resumes of all principals

    7.   Copies of letters of intent from suppliers, etc.

Business Plan: *Description of the Business*

In this section, provide a detailed description of your business. An excellent question to ask yourself is: "What business do I want to be in?" While answering this question, include your products, market, and services as well as a thorough description of what makes your bakery unique. Remember, as you develop your business plan, you may have to modify or revise your initial questions.

The business description section is divided into three primary sections: Section 1 actually describes your business; Section 2 states the product or service you will be offering; and Section 3 includes the location of your business and why this location is desirable (if you have a franchise, some franchisors will assist in site selection). The description of your business should clearly identify goals, objectives, and clarify why you are (or why you want to be) in business.

When describing your business, you should explain:

1. **Legalities.** Business form: proprietorship, partnership, or corporation; what licenses or permits you will need

2. **Business description.** 25-words or less describing business. An example would be: *Bakery specializing in elegant custom wedding cakes.*

   a. Memorize this quick pitch as you'll use it often with potential vendors and suppliers, bankers, lenders, and friends. Businesses aren't launched by the efforts of one person. Let people in on your dream and you'll find plenty of help.

3. **What your product or service is.** Perhaps a sample "daily offering" listing could be included.

4. **Business type.** Is it a new independent business, a takeover, an expansion, a franchise?

5. **Why your business will be profitable.** What are the growth opportunities? Will franchising impact on growth opportunities?

6. **When your business will be open.** What days? Hours?

7. **What you have learned about your kind of business** from outside sources (trade suppliers, bankers, other franchise owners, franchisor, publications).

A cover sheet goes before the description. It includes the name, address, and telephone number of the business and the names of all principals. In the description section, describe the unique aspects and how or why they will appeal to consumers.

Emphasize any special features that you feel will appeal to customers and explain how and why these features are appealing.

Legal Forms Of Business

When organizing a new business, one of the most important decisions is choosing the structure of the business. Your attorney and accountant can provide you with the legal and financial advantages and disadvantages of each. Factors influencing your decision about your business organization include:

- Legal restrictions
- Liabilities assumed
- Type of business operation
- Earnings distribution
- Capital needs
- Number of employees
- Tax advantages or disadvantages
- Length of business operation

Below you will find a brief description of the business entities you can choose:

Sole Proprietorship – The easiest and least costly way of starting a business
A sole proprietorship includes simply finding a location and opening the door for business. Start-up attorney's fees will be less than those of other business forms. The owner has absolute authority over all business decisions. The biggest negative to a sole proprietorship is your personal liability should the business default on a loan or be involved in a legal dispute.

Partnership – Two or more parties that join together to share ownership
The two most common partnership types are general and limited. A general partnership can be formed simply by an oral agreement between two or more persons, but a legal partnership agreement drawn up by an attorney is highly recommended. Legal fees for drawing up a partnership agreement are higher than those for a sole proprietorship, but may be lower than incorporating. A partnership agreement could be helpful in solving any disputes. However, partners are responsible for the other partner's business actions, as well as their own.

Corporation – A business entity where control depends upon stock ownership
A business may incorporate without an attorney, but legal advice is highly recommended. The corporate structure is usually the most complex and is more costly to

organize. Control depends on stock ownership. Persons with the largest stock owner-ship, not the total number of shareholders, control the corporation. Small, closely held corporations can operate more informally, but recordkeeping cannot be elimi-nated. Officers of a corporation can be liable to stockholders for improper actions. Liability is generally limited to stock ownership, except where fraud is involved. You may want to incorporate as a "C" or "S" corporation.

Limited Liability Company (LLC) – A blended form
The LLC is not a corporation, but it offers many of the same advantages. Many small business owners and entrepreneurs prefer an LLC because they combine the limited liability protection of a corporation with the "pass through" taxation of a sole propri-etorship or partnership. An LLC has advantages over corporations that allow greater flexibility in management and business organization.

Products/Services

Describe the benefits of your goods and services from your customers' perspective. Successful business owners know or at least have an idea of what their customers want or expect from them. This type of anticipation can be helpful in building customer satisfaction and loyalty. It certainly is a good strategy for beating the competition or retaining your competitiveness. Describe:

1. What you are selling. Include your bakery product listing and menu (if appropriate) here.

2. How your product or service will benefit the customer.

3. Which products/services are in demand (if there will be a steady flow of cash).

4. What is different about the product or service your business is offering. In marketing, this is called your USP (Unique Selling Position) and should be at the center of your marketing message.

The Location

The location of your business can play a decisive role in its success or failure. Remem-ber the old maxim, "Location, Location, Location"? Your location should be built around your customers, accessible, and provide a sense of security. Consider these questions when addressing this section of your business plan:

1. What are your location needs?

2. What kind of space will you need? Is there room to expand?

3. Are there environmental or zoning issues to be considered?

4. Why is the area desirable? The building desirable?

5. Is it easily accessible? Is public transportation available?
   Is street lighting adequate?

6. Is it affordable?

Think about Joining In

Notice how there seems to be a fast-food store on every corner. This isn't poor planning but a belief that dining and shopping means being within reach when the urge strikes. Being located near a green grocer, a fish market or a gourmet shop brings specific types of customers to your area. Donut or bagel shops benefit from traffic volume during peak sales periods.

To learn more about selecting your bakery location and how to build or buy a food service facility, research helpful articles on the web. This will give you a better idea of what you should pick and how to invest properly.

Location = Customers

After selecting various potential bakery sites or assessing existing bakeries for sale, be sure to obtain as many lifestyle and demographic facts as you can about each.

- How many bakeries of the same kind are located in the area?

- Can you find out something about their sales volume?

- Since bakeries attract primarily local inhabitants –
  - o What is the population of the area?
  - o Is the trend of population increasing, stationary or declining?
  - o Are the people native-born, mixed or chiefly foreign?
  - o What do they do for a living? Are they predominantly laborers, clerks, executives or retired persons? Are they all ages or principally old, middle-aged or young?

- To help you gauge your potential customers' buying power find out:
  - o The average sales price and rental rates for homes in the area

- o   Their per capita income
- o   Average family size

- Is the building/location suitable for a food service establishment? Zoning ordinances, parking availability, transportation facilities, and natural barriers—such as hills and bridges—are important in considering the environmental impact, assessability, and visibility of your bakery.

Business Plan – The Marketing Plan

Marketing plays a vital role in successful business ventures. How well you market your business, along with a few other considerations, will ultimately determine your degree of success or failure. The key element of a successful marketing plan is to know your customers—their likes, dislikes, and expectations. By identifying these factors, you can develop a marketing strategy that will allow you to arouse and fulfill their needs.

Who Are Your Customers?

Identify your customers by their age, sex, income/educational level, and residence. At first, target only those customers who are more likely to purchase your product or service. As your customer base expands, you may need to consider modifying the marketing plan to include other customers.

Develop a marketing plan for your business by answering these questions. (Potential franchise owners will have to use the marketing strategy the franchisor has developed.)

Customer Demographic Research

Your local Chamber of Commerce, state restaurant and bakery association, and peers can assist you with the demographics (economic and lifestyle patterns) of your community and business research.

Visit your local library or online databanks for additional data (ask about free research assistance from trained librarians):

- *Demographics USA (ZIP edition)* – market statistics

- *Lifestyle market analyst.* Standard Rate & Data Service – look under "gourmet cooking/fine foods" and cross-reference market, lifestyle, consumer.

- *Standard & Poor's Industry Surveys.*

Online check out:

- **www.prb.org**
- **http://quickfacts.census.gov**
- **www.searchbug.com/tools/demographics-data.aspx**

Your marketing plan should be included in your business plan and contain answers to the questions outlined below.

1. Who are your customers? Define your target market(s).

2. Are your markets (potential customer base) growing? Steady? Declining?

3. Is your market share growing? Steady? Declining?
   a. If a franchise, how is your market segmented?

4. Are your markets large enough to expand?

5. How will you attract, hold, and increase your market share? How will you promote your business?
   a. If a franchise, will the franchisor provide assistance in this area, based on the franchisor's strategy?

6. What pricing strategy have you devised?
   a. Remember price can be an important purchasing factor. Low prices aren't always the right pricing strategy. Focus on VALUE and SERVICE. Remember every grocery store in America sells bread; but you sell more than a loaf of flour and water.

Competition

An important part of determining your bakery's potential success is to study the competition. Finding your competitive edge or discovering a niche market is important. Learning what others are successful with can be helpful. In every business, keeping in tune with the competition is how you set yourself apart.

Questions like these can help you:

1. Who are your five nearest direct competitors?
2. Who are your indirect competitors?
3. How are their businesses: Steady? Increasing? Decreasing?
4. What have you learned from their operations? From their advertising?
5. What are their strengths and weaknesses?

6.   How do their products or services differ from yours?

Start a file on each of your competitors. Keep manila envelopes of their advertising and promotional materials and their pricing strategy techniques. Review these files periodically, determining when and how often they advertise, sponsor promotions ,and offer sales. Study the copy used in the advertising and promotional materials and their sales strategy. For example, is their copy short? Descriptive? Catchy? How much do they reduce prices for sales? Using this technique can help you to better understand your competitors and how they operate their businesses.

Scouting the Competition

This chapter has outlined some suggestions below to help you scout the competition. Remember that these are research expeditions to help you create your bakery's niche, learn from their experiences, and determine your potential for success.

- **Mark your potential location on a street map.** Draw a circle around the mark - for "walking" neighborhoods a 6-block radius should be enough; for "driving" neighborhoods start with a quarter mile. This is your initial study area. You can always expand your research territory based on how far you believe people will travel for your products.

- **Visit every business that serves baked goods "to-go" including grocery stores.** If you will be offering prepared take-out or dine-in items, you'll need to look at coffee shops, donut and bagel shops, delis, and cafés. *These are your competition.*

- **Sample your competition's products. Go to** establishments that feature similar menus, themes or serve your model customer.
  o   **Be a critic.** Make notes of what works and what doesn't.
  o   **Visit their retail outlet.** Even if they serve bagels, do they serve the same crowd?
  o   **Note how the patrons interact with the physical surroundings.** Are the lines confusing during peak hours? Are the displays attractive?

What Will You Sell Customers?

What you offer your customers every day (supplemented with seasonal offerings and daily specials) is called your product mix. Beyond baked goods, you may choose to have other "profit centers" such as take-out sandwiches, gourmet coffee, ice cream by the scoop or coffee accessories. Chapter 10 also discusses various food and non-food items that can comprise your product mix, how to satisfy customer demands, and how to focus on the profit-makers.

In your business plan, you'll discuss what products you will be offering, why you feel these are your best offerings for potential customers, and why they will buy them.

Pricing

Your pricing strategy is another marketing technique you can use to improve your overall competitiveness. Get a feel for the pricing strategy your competitors are using. That way you can determine if your prices are in line with competitors in your market area and if they are in line with industry averages.

Your bakery "style" and customer demographics are factors. Are you a family-friendly bakery with comfort foods? Do you cater to an executive crowd rushing off to work? Are your customers "foodies" seeking an elegant presentation or authentic French bread? Do your products appeal to the most discriminating buyers who appreciate your superior ingredients and handcrafting?

Some of the pricing considerations are:

- **Food cost** – Consumers understand that fresh and organic ingredients cost more.

- **Competitive position** – Mid-range pricing can be a great way to promote the value of your goods without moving into a more limited potential customer base.

- **Pricing below competition** – Undercutting the competition can bring people in, but it is quality and value that keeps them.

- **Pricing above competition** – Nancy Silverton's world-renowned La Brea Bakery has them lined up for $7 loafs of artisan bread. Elegant $3,000 wedding cakes wow guests across the nation.

- **Multiple pricing** – Offer multiple-buy discounts.

- **Service components** – Dine-in service means your pricing must cover a larger service staff.

- **Overhead costs** – Don't let your facility costs escalate your product pricing unreasonably. Sure it's great to be located in the new strip mall but how many $15 cakes will you have to sell every month just to meet expenses?

You'll find more information on pricing in Chapter 11.

Advertising and Public Relations

How you advertise and promote your bakery may make or break your business. Having a good product or service and not advertising and promoting is like not having a business at all. Many business owners operate under the mistaken concept that the business will promote itself and channel money that should be used for advertising and promotions to other areas of the business. Advertising and promotions are the lifeline of a business and should be treated as such. See Chapters 8 and 9 on marketing your bakery.

- Devise a plan that uses advertising and networking as a means to promote your business.

- Develop short, descriptive copy (text material) that clearly identifies your goods or services, your location, and price (remember price isn't always the deciding factor so this isn't always important).

- Use catchy phrases to arouse the interest of your readers, listeners or viewers.

- Make their mouths water! Don't just tell people what you sell – tell them why they should buy from you.

In the case of a franchise, the franchisor will provide advertising and promotional materials as part of the franchise package; you may need approval to use any materials that you and your staff develop. Whether or not this is the case, as a courtesy, allow the franchisor the opportunity to review, comment on, and, if required, approve these materials before using them. Make sure the advertisements you create are consistent with the image the franchisor is trying to project.

Your marketing plan section should include the various advertising and public relations efforts you will employ (how you will reach potential customers) and the cost of the campaigns. Assign percentages of your annual budget to the various media you will use. Don't forget to list any co-op advertising dollars you anticipate receiving from wholesalers, suppliers, and other businesses.

Other advertising efforts that should be included in your marketing plan are discount coupons, frequent buyer programs, point of sale displays, samples, and your open house.

Remember the more care and attention you devote to your marketing program, the more successful your business will be.

Business Plan – The Management Plan

Managing a business requires more than just the desire to be your own boss. It demands dedication, persistence, making decisions, and managing both employees and finances. Your management plan, along with your marketing and financial management plans, sets the foundation for and facilitates the success of your business.

Like a building and your baking equipment, people are resources—they are the most valuable assets a business has. You will soon discover that employees and staff (even if they are family members) will play an important role in the total operation of your business. Consequently, it's imperative that you know what skills you possess and what you lack since you will have to hire personnel to supply the skills that you lack.

It is imperative that you know how to manage and treat your employees. Make them a part of the team. Keep them informed of, and get their feedback regarding changes. Employees oftentimes have excellent ideas that can lead to new market areas, innovations to existing products or services or new product lines or services that can improve your overall competitiveness.

Your management plan should answer questions such as:

- How does your background/business experience help you in this business?
- What are your weaknesses and how can you compensate for them?
- Who will be on the management team?
- What are their strengths/weaknesses?
- What are their duties?
- Are these duties clearly defined?
- If a franchise, what type of assistance can you expect from the franchisor?
- Will this assistance be ongoing?
- What are your current personnel needs?
- What are your plans for hiring and training personnel?
- What salaries, benefits, vacations, and holidays will you offer? If a franchise, are these issues covered in the management package the franchisor will provide?
- What benefits, if any, can you afford at this point?

If a franchise, the operating procedures, manuals, and materials devised by the franchisor should be included in this section of the business plan. Study these documents carefully when writing your business plan and be sure to incorporate this material. The franchisor should assist you with managing your franchise. Take advantage of their expertise and develop a management plan that will ensure the success of your franchise and satisfy the needs and expectations of employees as well as those of the franchisor.

Business Plan – The Financial Management Plan

Sound financial management is one of the best ways for your business to remain profitable and solvent. How well you manage the finances of your business is the cornerstone of every successful business venture. Each year thousands of potentially successful businesses fail because of poor financial management. As a business owner, you will need to identify and implement policies that will lead to and ensure that you will meet your financial obligations.

To effectively manage your finances, plan a sound, realistic budget by determining the actual amount of money needed to open your business (start-up costs) and the amount needed to keep it open (operating costs). The first step to building a sound financial plan is to devise a start-up budget. Your start-up budget will usually include such one-time-only costs as major equipment, utility deposits, down payments, etc.

The start-up budget should allow for these expenses.

Operating Budget

An operating budget is prepared when you are actually ready to open for business. The operating budget will reflect your priorities in terms of how you spend your money, the expenses you will incur and how you will meet those expenses (income). Your operating budget also should include money to cover the first three to six months of operation. For more information on operating budgets, see Chapter 21.

The financial section of your business plan should include any loan applications you've filed, a capital equipment and supply list, balance sheet, break-even analysis, pro-forma income projections (profit and loss statement) and pro-forma cash flow. The income statement and cash flow projections should include a three-year summary (detail by month for the first year, then by quarter for the second and third years).

Developing projections are probably the most difficult aspect of writing your business plan. You should strive to be as accurate as possible based upon your research.

Overstating your sales in hopes of impressing lenders or investors can backfire on you should you not reach these estimates. Understating your sales can mean that you won't be prepared to produce sufficient enough volume to satisfy the demand.

The accounting system and the inventory control system that you will be using are generally addressed in this section of the business plan, also. Chapter 13 discusses computerized accounting and inventory systems.

If a franchise, the franchisor may stipulate in the franchise contract the type of accounting and inventory systems you may use. If this is the case, he or she should have a system already intact and you will be required to adopt this system. Whether you develop the accounting and inventory systems yourself, have an outside financial adviser develop the systems, or the franchisor provides these systems, you will need to acquire a thorough understanding of each segment and how it operates. Your financial adviser can assist you in developing this section of your business plan.

The following questions should help you determine the amount of start-up capital you will need to purchase and open a franchise.

- How much money do you have?
- How much money will you need to purchase the franchise?
- How much money will you need for start-up?
- How much money will you need to stay in business?

Other questions that you will need to consider are:

- What will your sales goals and profit goals for the coming year be? If a franchise, will the franchisor establish your sales and profit goals? Or will he or she expect you to reach and retain a certain sales level and profit margin?
- What financial projections will you need to include in your business plan?
- What kind of inventory control system will you use?

Your plan should include an explanation of all projections. Unless you are thoroughly familiar with financial statements, get help in preparing your cash flow and income statements and your balance sheet. Your aim is not to become a financial wizard, but to understand the financial tools well enough to gain their benefits. Your accountant or financial adviser can help you accomplish this goal.

Business Plan Resources

Business experts recommend that you write your own business plan, as the process is as important as the finished plan. There are numerous Web resources (search by "bakery business plan") to help you write a plan for internal and external purposes.

- **Purchase a fill-in-the-blank business plan book** from your local bookstore or a popular **software package** like *Business Plan Pro* from Palo Alto Software to get you started. *Business Plan Pro* includes a sample bakery business plan. These guides will ask thought-provoking questions and help you organize your thoughts, gather research data, and present it in a straightforward manner.

- **Check with your state's small business development center, the Small Business Administration (SBA) at <u>www.sba.gov</u> or your local university for free or low-cost business plan writing classes.** Besides learning how to write your plan, the discipline of attending a class can provide focus for busy entrepreneurs.

- **Hire a business communications specialist to "spruce up" your plan.** An experienced business plan writer can polish your presentation to dazzle bankers and investors.

Chapter 5

Launching Your Business – Pre-Opening Activities

• •

Each bakery launch has its own unique and challenging problems. Once you have set your desired opening date, you'll have plenty of pre-opening tasks to accomplish. Before engaging in any business activity, seek the guidance of a lawyer. You will undoubtedly have many legal questions and need legal counseling during the opening period. The services of a local accountant or CPA should also be retained. The accountant will be instrumental in setting up the business and can provide you with vital financial advice.

Governmental Laws, Regulations And Licenses

State Registration

Determine your state's requirements for starting a business as early as possible. A good place to start is your state's Secretary of State or Business Development Office website. All states have different regulations, so don't assume anything is final without verification.

There is generally a fee required for registering a new business. You will need to confirm that no other businesses are currently using your particular business name and file assumed business-name applications if required.

If your state has an income tax on wages, request from the State Department of Labor or Taxation all pertinent information. This would include all required forms, tax tables, and tax guides. Also, contact the State Department of Wage and Hour for their employee/employer regulations.

City Business License

Almost all cities and most counties require a permit to operate a business. Contact your local city hall for licensing information. The zoning board will check your application to see if your business conforms to all local regulations. Purchasing an existing bakery will eliminate most of these clearances.

Sales Tax

Contact the State Revenue or Taxation Agency concerning registry and collection procedures. Each state has its own methods of taxation on the sale of food products. Most states that require collection on food and beverage sales also require an advance deposit or bond to be posted against future taxes to be collected.

Sales tax is only collected on the retail price paid by the end user. Thus when purchasing raw food products to produce menu items it will not be necessary to pay sales tax on the wholesale amount. However, you must present the wholesaler with your sales tax permit or number when placing orders, and sign a tax release card for their files. Make certain that you comply with your state's regulations or you may be required to pay a use tax.

Certain counties and/or cities may also assess an additional sales tax in addition to the state sales tax. This entire issue needs to be thoroughly researched, as an audit in the future could present you with a considerable tax liability.

Health Department License

Your city's health department should be contacted as early as possible. A personal visit to discuss your plans and their needs should be in order. It would be to your advantage to show cooperation and compliance from the very beginning. The health department can and will close your facility until you comply with its regulations. A bakery shut down by the health department will almost surely be ruined if the closure becomes public knowledge. Before opening, the health department will inspect the premises. If the facilities pass the thorough inspection, they will issue the license enabling the bakery to open. The cost of the license is usually less than $50. Should they find faults in your facility, you will be required to have them corrected before they will issue a license.

Many states now have laws requiring that at least the manager—and in some states the entire staff—go through and pass an approved health and sanitation program. Check with your state bakery association on your state's specific requirements. The most common approved program is the ServSafe program developed by the NRA Educational Foundation. Materials may be purchased at **www.atlantic-pub.com** or by calling 800-541-1336. The ServSafe products, including instructor guides, answer sheets, instructor slides, manager training, food safety CD-ROMs and instructor tool kits, are available in English and Spanish.

Fire Department Permit

A permit from the fire department, which is also referred to as an occupational permit, is required before opening. The fire department inspectors will be interested in checking exhaust hoods, fire exits, extinguisher placements and the hood and sprinkler systems before you open.

Based upon the size of the building, the local and national fire code, and the number of exits, the fire inspectors will establish a "capacity number" of people permitted in the building at one time. Follow their guidelines strictly, even if this means turning away customers because you've reached capacity.

Building and Construction Permit

Should you plan on doing any renovating to the bakery that is going to change the structural nature of the building, you may need a local building permit. Permit approval for retrofitting a building not previously used for cooking/baking may take time, so apply early.

Building permits are generally issued from the local building and zoning board. You will need to contact the building inspector with your blueprints or plans to initially determine if a permit is required. Should a permit be required he or she will inspect your plans, ensuring that they meet all the local and federal ordinances and codes.

Once the plans are approved, a building permit will be issued. The building inspector will make periodic inspections of your work at various stages of completion to ensure that the actual construction is conforming to the approved plans.

Sign Permits

Many local governments are beginning to institute sign ordinances and restrictions. These ordinances restrict the sign's size, type, location, lighting, and the proximity of the sign to the business. The owners or managers of a shopping mall or shopping center may also further restrict the use and placement of signs.

Federal Identification Number

All employers, partnerships and corporations must have a Federal Identification Number. This number will be used to identify the business on all tax forms and other licenses. To obtain a federal identification number fill out Form #554, obtainable from the IRS, use their web registration at **www.irs.gov** or enroll by phone by calling 800-829-4933. There is no charge.

Opening the Bakery Bank Account

Opening a business bank account is a great deal more important than at first glance. If you received your financing through a local commercial bank, you should also use this bank for your business account, if it fills all your needs.

Take plenty of time to shop around for the bank that will serve you the best. Many smaller banks are better suited for personalized service. When you go into a prospective bank, ask to see the bank manager, tell him or her of your plans and discuss what your needs are. Look at what each bank charges for check and deposit transactions and all other service charges. Compare credit-card fees. Even a small percentage can add up to a great deal of money over the years.

Insurance

Properly insuring a bakery is similar to the coverage of any business enterprise where members of the public are in frequent attendance. Liability protection is of the utmost concern. Product liability is also desirable, as the consumption of food and beverages always presents a hazard. At the very least, your business will need fire, liability and, workers' compensation insurance. A discussion with your agent will help determine which insurance policies you should purchase. Your franchisor, lessor, mortgage holder or investors may have specific insurance requirements that need to be taken into account also.

Workers' Compensation Insurance

Workers' compensation insurance covers loss due to statutory liability as a result of on-the-job personal injury or the death of an employee. Some states also provide for stress-related claims. This insurance coverage pays all medical treatment and costs plus a percentage of the employee's salary due to missed time resulting from the injury. Workers' compensation insurance is mandatory in most states. Failure to comply can result in large penalties and fines. Be certain to obtain all the information that pertains to your particular state.

Organizing the Pre-Opening Activities

Opening a bakery or any business is a great test of anyone's organizational and managerial abilities. It is imperative that communication be maintained with your key personnel. Keep track of the assignments that need to be completed, who the assignments are delegated to, and when they must be completed. Allow plenty of time for assignments and projects to be accomplished. Even the seemingly simplest task may uncover a web of tangles and delays. Delegate responsibilities whenever possible, but above all else, keep organized. Maintain a collective composure and deal with people and problems on a level and consistent basis and you'll be off to a great start.

Pre-Opening Promotion

Described below are some pre-opening promotional ideas. It should be noted that there is a definite distinction between promotion and advertising. Promotion involves creating an interest in a new project usually at little or no cost. You'll find additional marketing and promotional information in Chapters 8 and 9.

As soon as possible, put up the new bakery sign or a temporary sign explaining briefly the name of the new bakery, specialties, hours of operation, and the opening date. People by nature are curious about what is occurring in their neighborhood; give them something to start talking about. This is perhaps the best and least expensive promotion you can do.

1. Meet with the advertising representatives for the local papers. Determine advertising costs and look into getting a small news story published describing the bakery.

2. Have plenty of business cards on hand as soon as possible: they're a great source of publicity. Your business card can double as a frequent-buyer card or have a special manager discount printed on the back to bring people in.

3. Join the Better Business Bureau and the local Chamber of Commerce. Besides lending credibility to your organization, they often can supply you with some very good free publicity.

4. When you place your employment ad in the classified section always list the type of bakery and location. This inexpensive classified advertising will help spread the word. Many people in the food service industry also love to find new bakeries to try out.

5. Plan a grand opening celebration! That's an excellent reason for the local news to feature you. Create grand opening advertising such as a door hanger campaign or your first newspaper ad with a discount coupon.

Contacting Purveyors and Supplies

Approximately six to eight weeks before the scheduled opening date, it will be necessary to contact all the local suppliers and meet with their sales representatives. It would be advisable to have your bakery manager and other key personnel attend.

Opening Labor

Before the opening date there may be many people on the payroll. You will need the assistance of personnel to assemble chairs, decorate, test recipes and equipment, and do anything required so that the opening date may be achieved. Many of these temporary employees may be utilized for various jobs in the bakery after opening. The time clock should certainly be used during this period for better control. Overtime must be carefully monitored, and if at all possible, avoided. This will require a great deal of organization between assignments and scheduling. To learn more about hiring employees and payroll activities, see Chapters 6 and 20.

Public Utilities

Notify public utility companies of your intention to be operating by a certain date. Allow plenty of lead-time for completion.

Phone Company

You will need at least two phone lines for any bakery. Don't lose customers because they can't get through. You should have two phones in the offices, and one to two extensions at the register or order-taking area. Place local emergency numbers at all phones.

Gas and Electric Companies

All major equipment needs special hookups that can only be completed by trained technicians of either the gas or electric company (or their authorized representatives) or licensed equipment installers (electricians, plumbers or factory support personnel). They should be contacted as early as possible to evaluate the amount of work required. In many cases, they will need to schedule work several weeks ahead of time.

Water

Water is different in all parts of the country due to the type of chemical particles it contains. Water that has been subjected to a chemical treatment plant may contain a high level of chlorine. Water taken directly from the water table will contain any number of additives depending upon the geological makeup of the soil where it came from. Different types of water can give different results when used in baking. Your local water district can give you information concerning the water's chemical makeup.

Chemical particles in the water can have a particularly bad effect in the brewing of fresh coffee. Water quality and mineral composition can affect baked goods and other foods using water. Excessively hard water is undesirable when baking yeast products as it inhibits fermentation. A commercial water softener may be necessary to reduce the pH.

Several companies now have on the market filtering devices that attach directly to the water lines. If prescribed, filters need only be connected to the water lines that are used for drinking/cooking water. Bathroom and dishwasher lines would not require a filter. Filtering devices are usually tube-shaped canisters that contain charcoal or a special filtering paper. Discuss your particular situation with your state Department of Natural Resources and the sales representative for your coffee supplier.

Security Needs

Locksmith

A registered or certified locksmith can rekey locks as soon as you occupy the building. Keys to locked areas should be issued on a need-to-have basis. The locksmith can set door locks so that certain keys may open some doors, but not others. Only the owner and manager should have a master key to open every door. Each key will have its own identification number and "Do Not Duplicate" stamped on it. Should there be a security breach, you can easily see who had access to that particular area. The bakery should be entirely rekeyed when key-holding personnel leave or someone loses his or her keys. Safe combinations should be periodically changed as well.

Fire and Intrusion Alarms

Every bakery should have a system for fire, smoke and heat detection and, depending upon your location, for intrusion and holdup. Check with your fire marshal for your bakery's fire detection and suppression requirements. They may also direct you towards a reputable fire and safety service company.

The fire detection system consists of smoke monitors and heat sensors, strategically placed around the building. This system must be audible for evacuations and directly connected to either the fire department or a private company with 24-hour monitoring service. In newer buildings, the sensors also activate the sprinkler system.

Most cities and states also require food service establishments to install a hood system in the cooking areas. This consists of a sprinkler-type system situated above equipment with an exposed cooking surface or flame. The system may be operated either automatically or manually. When released, a chemical foam immediately sprays out over the area. This is particularly effective in stopping grease fires. Once activated the system will automatically shut off the gas or electric service to the equipment. In order to regain service, the company that installed the system must reset it.

Depending upon your community, the installation of an alarm system in the bakery is almost a necessity. The loss of business and profits due to burglary, vandalism or arson is not to be gambled upon. The installation of an alarm system will increase the value of the property, and a 24-hour monitored system may make you eligible with your insurance company for a rate reduction of 5 to 10 percent on the insurance premium.

Another recommended option to obtain from a private monitoring service is overseeing the temperature in freezers and walk-ins. A temperature-sensitive device may be installed in the freezer and walk-ins. When the temperature rises to a certain level, it will set off a monitoring station alarm. The operator may then call the bakery manager or dispatch the refrigeration repairman. Some security service companies may also provide other services including guard service in the lounge area and escort service to the bank. These companies must of course be bonded, licensed, and insured.

Support Service Providers

Dishwasher Chemical Company

Contact all the dishwasher chemical suppliers in the area and meet with their representatives. In most areas, you'll have four or five companies that supply chemicals only and some that rent dishwashers (service is included) to chemical buyers. These companies maintain research staffs that are constantly developing innovative chemicals and devices that will lower the machine's energy and chemical usage and your operational cost. Their field service personnel are experts capable of monitoring the entire system and guiding your staff on how to operate it efficiently. Clean dishware is an absolute necessity for a bakery. Don't gamble on the outcome by not using an expert.

Knife Sharpener

The services of a professional knife sharpener should be utilized in any bakery. Sharp knives, scissors and other cutting instruments are essential for safety (a dull knife is more dangerous than a sharp one) and productivity. Generally, a service contract may be purchased so that all knives and the blades of the cutting and slicing machines will be sharpened on a regular basis. In between servicing, the staff may keep the blades honed on a sharpening oil stone or ceramic sharpening sword.

Sanitation Service

A bakery of any size has a great deal of waste. In order to preserve a proper health environment, the services of a trash removal or sanitation service company will be required. Waste and recycling issues are detailed in Chapter 15.

If your local municipality or shopping center/mall owner does not provide trash removal, gather quotes from all the sanitation companies in the area. Prices may vary considerably depending upon who purchases the dumpsters. You may wish to get advice from your health department for the selection. Any service contract should contain provisions for the following:

- Dumpsters with locking tops.

- Periodic steam cleaning of the dumpsters.
- Fly pesticide sprayed on the inside of the dumpster.
- Number of days for pickup.
- Extra pickups for peak periods.

Parking Lot Maintenance

Many mall/shopping centers include parking maintenance in your lease. If it is not noted, negotiate this, as sharing the cost with other merchants makes the most sense. If you are responsible for your own parking area, remember that parking lots will need periodic maintenance besides the daily duty of light sweeping and picking up any trash. Painting new lines for the parking spaces should be done annually. Blacktop surfaces will also require a sealant to be spread over the surface periodically. This stops water from seeping into it. Winter climates will require snow removal and possibly salting and sanding of the lot. Most of these services may be purchased under contract.

Plumber

A local plumber will be needed to handle any miscellaneous work and emergencies that may come up. The plumber must have 24-hour emergency service. Make every effort possible to retain the plumber that did the original work on the building. He will be thoroughly familiar with the plumbing and know why certain procedures were performed. This can be a terrific advantage.

Electrician

As with the plumber, it would be a great advantage to retain the original electrician who worked on the building. An electrician will be needed when equipment is moved or installed. If it has not been done already, the electrician should check out and label all the circuits and breakers in the building. The electrician should also be on 24-hour emergency service.

Refrigeration Service

The most important consideration when choosing a refrigeration company is how fast they can respond to emergencies. At any given time the refrigeration systems and freezer could go out, which may result in the loss of several thousand dollars in food. Make certain any prospective company understands this crucial point. They must have 24-hour service.

Exterminator

Exterminators must be licensed professionals with references from the other food service businesses they service. Extermination methods should comply with local health department regulations and be guaranteed food safe. Exterminators can eliminate any pest-control problems, such as rats, cockroaches, ants, termites, flies, etc. Have several companies come in to appraise the building. They are experts and can read the "telltale" signs that might otherwise be missed. The company selected should be signed to a service contract as soon as possible. This is not an area to cut corners or try to do it yourself—it won't pay in the long run.

Plant/Landscape Maintenance

A professional plant-care and/or landscape person can provide all the necessary services to plant, rotate, water, transplant, and arrange indoor and outdoor plant decor. Contact the companies in the area and get their opinions, quotes, and references. They must be made aware that they are working in an environment where toxic sprays may only be used with the approval of the health department, and even then very cautiously.

Heating, Ventilation and Air-Conditioning

You will need the services of a HVAC company that can respond 24 hours a day at a moment's notice. Make certain the company is reliable with many references.

Losing the heating system in the winter or the air-conditioning in the summer can put undo strain on your staff and customers. Providing workers with a cool place to rest and relax away from overheated baking areas will improve their efficiency.

Depending upon your community's air quality standards, an air filtration system may be required. VOC (volatile organic compounds), particulates and ammonia emissions are common in bakeries. Contact your state air quality control division for information on air filtration and emission control systems required in your area. Your heating/cooling service company should be able to assist you with your filtration needs.

Exhaust Hood Cleaning Service

Contact a company that specializes in the cleaning of exhaust hoods and ventilation systems. They should appraise and inspect the whole ventilation system before opening. Depending upon the amount and type of cooking performed, they will recommend a service that will keep the system free from grease and carbon buildup.

Maintaining your ventilation system will help it run more efficiently, reduce dust, and conserve energy. Usually twice-a-year cleaning is required. Many of these companies also offer grease and fat (deep fryer oil) removal.

Janitorial and Maintenance Service

Depending upon the size and operating hours of the bakery, you may wish to use the services of a professional cleaning company with experience in food service establishments. Cleanliness has an important effect upon your employee morale. A spotless bakery will create the environment for positive employee work habits. Outsourcing should be for cleaning – not sanitizing, which is your staff's responsibility.

The maintenance service company selected must have impeccable references. The company should be insured against liability and employee stealing. Employees should be bonded. You will probably need to give the owner of the company his or her own keys to the entrance, maintenance closets, security system, and possibly the office, for cleaning. It must be made very clear that food is completely off limits to their employees.

Some equipment manufacturers will provide detailed cleaning instructions. Special cleaners must be used on some equipment. Improperly cleaning a piece of equipment can ruin it forever. Keep this information in a loose-leaf binder in the office. The cleaning supervisor should have access to this manual and must be thoroughly familiar with its contents.

Coffee Service Vendor

Major coffee distributors offer the same basic plan for serving your customers coffee. They will provide—at no charge—the equipment necessary for coffee service including: brewing machines, filters, vacuum pots, serving pots, and maintenance and installation of all equipment. All that is required from you is to sign a contract stating you will buy their coffee exclusively. The price of all the equipment and maintenance is included in the price of the coffee.

When negotiating with the coffee salesperson, inform him or her that you want brand-new equipment. They are competing for your business, but once you sign the contract, you will be locked in to it. Use this leverage now while you have it. Placing the coffee machines in the main and service bars for the making of coffee drinks would increase efficiency greatly.

Various specialty teas may be purchased from these distributors. Sugar packets and sugar substitute packages may also be purchased from these companies. For an additional charge, your bakery name and logo can be imprinted on the outside of each packet and on disposable insulated cups and sleeves.

Soda Systems Vendors

Soda systems may be contracted in the same manner as the coffee arrangement. National-brand soda distributors will connect all the hoses, valves, taps, and guns needed to operate the bar, generally at no cost. You will be obligated to sign a contract stating that you will purchase their products exclusively. The price of the system is passed on to you as you purchase soda canisters. The distributor will also provide promotional material such as wall plaques, neon lights, drink coasters, etc.

These systems do occasionally break down, so maintain at least two cases of each type of soda in bottles or cans. Soda lines must be flushed out every week. The cost of this service is usually your responsibility. You may do this yourself, and many people do; however, it is recommended that the services of a professional be used. They are experts and have the proper equipment to do the job thoroughly. Cleaning these lines is not something that should be experimented with. The distributors can recommend a service.

Florist

Should you decide to have fresh cut flowers, you will need to contact a local florist. Each week the florist will deliver cut flowers or plants of your choice. Many dining rooms use only a single flower or rose in a long-stem vase with some fern for backing.

Linen/Uniform Service

Most bakeries won't require table linens but should you decide to kick it up a notch for customers, a linen service can make the process easier. Linen service companies can also provide food service uniforms. Fresh clean aprons and uniforms are a must. Compare and explore the cost of this service with purchasing an adequate supply that would be laundered in-house or by employees.

Chapter 6

Successful Employee Relations and Labor Cost Control

The most important investment you'll ever make for your business is in the people who work for and with you. No customer will return, no equipment will save time, and no cookies will melt in your mouth without your employees. From the janitor to the master baker, every person plays an important role in your success. Even in the smallest bakery, you cannot do it alone. Having great employees starts with being a great boss.

The Value (and Cost) of Employees

Employees directly control food quality and presentation. A disgruntled employee will not produce or perform as well as a satisfied one. Yet, it is bewildering to note that many owner/managers fail to make the employee's job any easier or more enjoyable with modern training procedures.

You must make every conceivable effort to relieve employees of "busy" work and make their jobs more rewarding—the results can only be positive. Unfortunately, most operators cannot substantiate or rationalize any investment in an employee's comfort or training because they feel the employee will probably quit in a month or two; they think, *why bother?* Of course, this starts the cycle all over again.

Take the initiative to resolve the problem, because you are the only one who can! Even simple accommodations—such as separate employee lockers, restrooms, and break rooms— can make all the difference. Food prep conditions are often harsh; air conditioning in the hot kitchens is virtually nonexistent; light and noise levels are usually inadequate, causing fatigue, inefficiency and accidents.

All of these poor physical conditions have been resolved in the manufacturing industry, where the turnover rate is around 8 percent and employees remain for several years or, as in Japan, where employees are virtually guaranteed economic security for their entire lifetimes.

Employees are one of the greatest resources a bakery has, but this resource will be wasted if you do not first recognize it, and then supply the proper incentive and motivation necessary to harvest it. You must provide employees with:

- Higher salaries
- Better training
- Insurance programs
- Flexible scheduling
- Shorter workweeks
- Childcare and transportation vouchers
- Incentive plans
- Proper tools to increase productivity, reduce stress and maximize safety
- Safe, clean working conditions
- Proper training and evaluations
- Financial security if possible
- Adequate benefit packages
- An opportunity for advancement
- An amicable, structured, and just working environment

The cost of providing these basic necessities, which have eluded the food service industry for so long, can be easily substantiated when compared to the ramifications of losing—and the cost of replacing—a discontented skilled employee.

First, consider the indirect cost to the bakery when an unmotivated and unhappy employee prepares a food item or provides poor service to a regular customer. Just as word-of-mouth advertising can be great publicity, it can also cause the bakery to come crumbling to the ground from comments like *"The food just was as good as it usually is,"* or *"The cake was wonderful, but the cashier was rude. I can get the same thing down the road for much less."*

Discontented employees will not be concerned about looking out for the bakery's interests. Dishes will be broken or carelessly thrown away. Food costs will rise and work areas will be left unorganized and dirty. Why should employees care about the bakery's profits when they barely make enough to survive?

Consider the direct cost of replacing an employee: recruiting expense, interviewer's salary and time, administrative cost, training expense, medical exams, the loss of sales and the cost of materials due to training mistakes, the labor cost paid before the employee's full productivity is reached, and the trainer's and supervisor's salaries. Consider the cost of termination: paperwork, exit interviewing and, possibly, unemployment compensation. According to the American Management Association, the cost to replace an employee who leaves is, conservatively, 30 percent of his annual salary. For those with skills in high demand, the cost can rise to a frightening 1.5 times the annual salary. Your ability to retain the kind of workers you want and need has a direct impact on the profitability and effectiveness of your organization.

Bakery operators must recognize the importance of employees to their success and take the necessary steps to ensure their physical comfort and economic security. Employee relations are an area where corners cannot be cut; the long-term results will outweigh the initial cost. Competent skilled labor is a finite resource with every business competing in the marketplace.

Good employees want to work for people who appreciate their skills and can provide them with the proper compensation for their efforts. The food-service industry must bring its labor policies and procedures up to date before it loses its personnel to other industries or is forced to change by union organization. Your business model must include paying employees a living wage, providing benefits, and respecting their talents and ideas.

The following sections will describe in detail, from the initial interview through to the employee's termination, how to set up and administer admirable manager/employee relations for the mutual benefit of each party.

Hiring Bakery Employees

The key to hiring good, competent employees is to put aside personal prejudices and select one applicant over another only because you feel he or she will have a better chance of being successful at the job. What a potential employee is qualified and capable of doing is often quite different from what he or she actually will do. The purpose of this section is to provide the interviewer with the information necessary to determine if the applicant has the qualities needed.

The Food-Service Labor Statistics: Industry on the Rise

According to the National Restaurant Association, the food-service industry employs 14 million people today, which represents 10 percent of the total workforce. By the end of 2015, the restaurant industry will have made about $709.2 billion across the U.S. Although the food-service industry declined back in the early 2000s, its revenue has increased every year since and will continue to do so.

The shortage of workers is not unique to the labor side of the equation; good management is also hard to find. The turnover in the population of senior food service managers and bakery artists isn't likely to slow down any time soon. It's very important to study

The country's tight labor conditions are forcing food-service companies to seek out new sources of labor: minority groups, welfare recipients, senior citizens, and the physically handicapped. As a result, the workforce is becoming more diverse. In addition, food service operators are extending their recruiting efforts beyond traditional methods (such as newspaper ads or signs posted in the bakery) to high school and college campus recruitment, retirement communities, state agencies, and the Web.

Key Points For Conducting Employment Interviews

A.   Treat all applicants considerately and show a genuine interest in them, even if they have little or no chance of obtaining the job. Every applicant should be treated as a potential employee, because they are.

B.   Make certain that you are on time and ready to receive the applicant. Arriving late or changing appointment dates at the last minute will give the applicant the impression that you are unorganized and that the bakery is run in the same manner.

C. Know the job being offered thoroughly. You cannot possibly match someone's abilities with a job you do not know or understand completely.

D. All interviews must be conducted in privacy, preferably in the interviewer's office. Interruptions must be kept to a minimum.

E. Make the applicant feel at ease. Have comfortable chairs and possibly beverages available. Speak in a conversational, interested tone.

F. Applicants will be full of questions about the job, its duties, the salary, etc. Newspaper advertisements tell only a little about the job and your company, so allow plenty of time for this important discussion.

G. Whenever possible, let the applicants speak. You can learn a great deal about them by the way they talk about themselves, past jobs, former supervisors, and school experiences. Watch for contradictions, excuses and, especially, the applicant being on the defensive or speaking in a negative manner. Avoiding subjects is an obvious indication that there was some sort of problem there in the past; be persistent about getting the whole story, but don't be overbearing. Come back to it later if necessary.

H. Never reveal that you may disapprove of something an applicant has done or said; always appear open minded. On the other hand, don't condone or approve of anything that is obviously in error.

I. Always ask a few questions they don't expect and aren't prepared for: What do they do to relax? What are their hobbies? What is the last book they read? Try to understand their attitudes, personalities, and energy levels.

J. Perhaps one of the most useful things you can ask when interviewing prospective employees is, "What were your favorite parts of your previous job?" Look to see if the things they liked to do with previous employers fit with the things you'll be asking them to do for you. It is important to cross-train employees to do as many jobs as possible, and it helps us to know which of those jobs will be a good fit.

    a. Often in interviewing prospective food service employees you'll get two types of applicants—those who say they prefer the "people part" of the job (talking to customers, serving customers, running the cash register) and those who like the "food part" of the job (bakery, food prep, short order cook). Most applicants will be honest about what they like to do.

K. Be sure to ask at least one behavior-based question. This will be very useful in getting at how an applicant responds in real-life work situations and how well he or she is able to handle them. For example: "What would you do if a customer complained that the 'whip cream just doesn't taste right'?" Or, "What would you do if your seemingly happy patron did not leave any tip at all?"

Unlawful Pre-Employment Questions

This section is not intended to serve on behalf of, or as a substitute for, legal consul, or even as an interpretation of the various federal and state laws regarding equal and fair employment practices. The purpose of this section is only to act as a guide to the type of questions that may and may not be legally asked of a potential employee. For more information on the topic, contact your state labor board.

A discussion of this subject with both the state and federal labor offices and with your lawyer would be in order. Standard employment applications may be purchased at your office-supply store. Before you use these forms, let your lawyer examine one to make certain that it doesn't contain or insinuate any questions that might be considered illegal.

The Federal Civil Rights Act of 1964 and other state and federal laws ensure that a job applicant will be treated fairly and on an equal basis, regardless of race, color, religious creed, age, sex or national origin.

In order to support these regulations, you cannot ask certain questions of applicants concerning the aforementioned categories. There is a fine line between what may and may not be asked of applicants. Use basic common sense concerning the type of questions you ask. Any illegal question would have no bearing on the outcome of the interview anyway, so avoid questions that are related to, or might evoke an answer that infringes upon, the applicant's civil rights.

- **Age** is a sensitive pre-employment question, because the Age Discrimination in Employment Act (on the Web at **http://eeoc.gov/laws/statutes/adea.cfm**) protects employees 40 years old and above.

  o It is permissible to ask an applicant to state his or her age if it is fewer than 18 years to determine if they will require a state-issued work permit. The U.S. Department of Labor provides information on child labor laws at **www.dol.gov/dol/topic/youthlabor/index.htm** as can your State Labor Board.

o   If you need the date of birth for internal reasons, e.g., computations with respect to a pension or profit-sharing plan, this information can be obtained after the person is hired.

* **Drugs, smoking:** It is permissible to ask an applicant if he or she uses drugs or smokes. The application also affords an employer the opportunity to obtain the applicant's agreement to be bound by the employer's drug and smoking policies. The application also affords an employer an opportunity to obtain the applicant's agreement to submit to drug testing.

* **Other problem areas:** Questions concerning…

    o   Whether an applicant has friends or relatives working for the employer may be improper if the employer gives preference to such applicants.

    o   Credit rating or credit references have been ruled discriminatory against minorities and women.

    o   Whether an applicant owns a home have been held to be discriminatory against minority members, since a greater number of minority members do not own their own homes.

    o   While questions about military experience or training are permissible, questions concerning the type of discharge received by an applicant have been held to be improper, because a high proportion of other-than-honorable discharges are given to minorities.

    o   The Americans with Disabilities Act prohibits general inquiries about disabilities, health problems and medical conditions. The U.S. Department of Labor offers ADA information for employers at **www.dol.gov/odep.**

A list of prohibited questions, some of which are obvious but used to illustrate the point:

* How tall are you?
* What color are your eyes?
* Do you work out at the gym regularly?

- Do you or anyone you know have HIV?
- Did you get any workers' comp from your last employer?
- How old are you?
- Have you been in prison?
- Are you really a man?
- Do you rent or own your home?
- Have you ever declared bankruptcy?
- What part of the world are your parents from?
- Are you a minority?
- Is English your first language?
- I can't tell if you're Japanese or Chinese. Which is it?
- So which church do you go to?
- Who will take care of the kids if you get this job?
- Is this your second marriage, then?
- Just curious: Are you gay?
- Are you in a committed relationship right now?
- How does your boyfriend feel about you working here?

Screening Potential Employees

Screening job applicants will enable you to reject those candidates who are obviously unsuitable before you schedule a lengthy interview. This saves both the bakery's and the applicant's time and money. An assistant manager, or someone else with background knowledge about your bakery, can conduct preliminary screenings. Potential job candidates may then be referred to the manager for intensive interviews.

All applicants should leave feeling they have been treated fairly and had an equal opportunity to present their case for getting the job. As previously stated, this is an important part of public relations.

Base your preliminary screening on the following criteria:

1. **Experience**—Is the applicant qualified to do the job? Examine past job experience. Check all references.

2. **Appearance**—Is the applicant neatly dressed? Remember he or she will be dealing with the public. The way the applicant is dressed now is probably better than the way she will come to work.

3. **Personality**—Does the applicant have a personality that will complement the other employees' and impress customers? Is he or she outgoing but not overbearing?

4. **Legality**—Does the applicant meet the legal requirements?

5. **Availability**—Can the applicant work the hours needed? Commute easily?

6. **Health and physical ability**—Is the applicant capable of doing the physical work required? All employees hired should be subject to approval only after a complete physical examination by a mutually approved doctor.

Make certain the application is signed and dated.

All applicants at this point should be divided into one of the three following categories:

1. **Interview applicant**— If you are not the decision-maker in hiring, refer the applicant to the manager and/supervisor to schedule an interview. Involve others who will be responsible for the new hire. You'll show your management team that their opinions are important and you'll improve your hiring practices.

2. **Reject**—Describe the reasons for rejection and place the application on file for future reference. NEVER make written comments that are in any way derogatory as described in the discrimination areas above. Check with your lawyer on recommended retention times for these applications.

3. **Prospective file**—Any applicant who is not qualified for this position but may be considered for other areas of the bakery should be placed in a prospective applicant file.

Once you have categorized all prospective candidates, be courteous and let those in category #2 and #3 that they have not been selected for the position. For those in the "future prospect" category, ask permission to contact them when another opening arises.

What to Look for in Potential Bakery Employees

1. **Stability**—You don't want employees to leave in two months. Look at past employment dates. Stability also refers to the applicant's emotional makeup.

2. **Leadership qualities**—Employees must be those who are achievers and doers, not individuals who have to be led around by the hand. Look at past employment positions and growth rate.

3. **Motivation**—Why is the applicant applying to this bakery? Why the bakery industry in general? Is the decision career related or temporary? Does the applicant appear to receive her motivation from within or by a domineering other, such as a spouse or parent?

4. **Independence**—Is the applicant on his own? Does he appear to be financially secure? At what age did he leave home? And for what reasons?

5. **Maturity**—Is the individual mentally mature enough to work in a stressful environment? Will she be able to relate and communicate with other employees and customers who may be much older than her?

6. **Determination**—Does the applicant seem to always finish what he starts? Does he seem to look for, or retreat from, challenges? Examine time at school and at last job.

7. **Work Habits**—Is the applicant aware of the physical work involved in bakery employment? Has the applicant done similar work? Does the applicant appear neat and organized? Look over the application; is it filled out per the instructions? Neatly? In ink? Examine past jobs for number and rate of promotions and raises.

The Final Selection

Reaching the final selection, to whom to offer the position, is often a difficult choice. You may have more than one applicants who are qualified and would probably become excellent employees, but which one should you decide upon?

Always base your choice on the total picture the applicants have painted of themselves through interviews, resumes, and applications. Gather advice from those who interviewed or had contact with the individuals. Not only will this help you reach the correct decision but will also make the rest of your staff feel a part of the management decision-making team. Whomever you select, he or she must be someone you feel good about having around, someone you hopefully will enjoy working with, and whom you feel will have a very good chance of being successful at the job.

Once you offer him or her the job, make certain the applicant fully understands the following items before accepting it:

1. **Salary** – Starting pay, salary range, expected growth rate, the day payroll is issued, company benefits, vacations, insurance, etc.

2. **Job description** – List of job duties, hours, expectations, etc.

3. **Report date** – Time and date of first day of work and to whom he/she should report.

4. **Job attire** – You probably talked about this during the interview but make certain they understand your rules about facial hair, jangling bracelets, and other appearance issues BEFORE they start. Emphasize how they relate to good customer service and employee safety.

5. **Drug testing** – Clarify the procedures and let them know that their employment would not begin before the testing is completed with negative results.

Rejecting Applicants

Rejecting applicants is always an unpleasant and difficult task. The majority of the applications will be rejected almost immediately. Some applicants will ask the reason for rejection. Always be honest, but use tact in explaining the reasoning behind the decision. Avoid a confrontation, explaining only what is necessary to settle the applicant's questions. Usually it will be sufficient to say, "We accepted an applicant who was more experienced," or "…who is better qualified."

As mentioned before, some applications may be transferred into a "prospective file" for later reference. Inform the applicant of this action, but don't give the impression that he or she has a good chance of being hired, nor state a specific date when you will be looking for new employees.

Employee Handbook/Personnel Policy Manual

Federal law mandates that all employers, regardless of size, have written policy guidelines. Employee handbooks/policy manuals are used to familiarize new employees with company policies and procedures. They also serve as guides to management personnel.

Formally writing down your policies could keep you out of lawsuits, prevent problems and misunderstandings, save time spent answering common questions, and look more professional to your employees. Explaining and documenting company policy to your employees has been proven to increase productivity, compliance, and retention.

Lack of communication, along with inadequate policies and guidelines, has been cited as a major factor in workplace legal disputes. Failure to inform or notify employees of standard policies has resulted in the loss of millions of dollars in legal judgments. Simply not being aware that their actions violated company policy has been an effective defense for many terminated employees. Most importantly, have the employee sign a document stating he or she has received and reviewed the handbook, and understands and intends to comply with all policies in the manual.

Writing Your Employee Policy Manual

If you have ever written a policy document before, you know how time consuming it can be. Even if you were a lawyer, it would likely take you 40 hours to research and write a comprehensive employee manual.

To pay someone to draw one up for you can cost thousands of dollars. Atlantic Publishing has put together a standard employee handbook guide for the food-service industry. All you have to do is edit the information. The template contains all of the most important company handbook sections and is written in Microsoft Word, so customizing and printing your manual is as easy as possible. The program currently sells for around $70 and is available at **www.atlantic-pub.com** or by calling 800-541-1336 (Item EHB-CS). There is also an order form at the back of this book.

There are a variety of human resource websites that can assist you with writing a custom employee manual or you can purchase a standardized form. Your attorney can also guide you through this process and can read your final version for compliance with state and federal laws.

Personnel File

Once the applicant is hired, an individual personnel file should be set up immediately. It should contain the following information:

- Application.
- Form W-4 and social security number.

- Name, address, and phone number.
- Emergency contact information and phone number.
- Employment date.
- Job title and pay rate.
- Past performance evaluations.
- Signed form indicating receipt and acceptance of employee handbook/ personnel policy manual
- Termination date, if applicable, and a detailed account of the reasons for termination.

Employees may also elect to have medical alert information (such as allergies) in their files for emergency situations.

Training

The most serious problem facing labor relations in the food service industry today is the lack of trained personnel and structured, industry-wide training programs. New bakery employees are often thrown into jobs, with little or no formal training. While on the job they must gather whatever information and skills—whether correct or not—they can. Blame for this situation lies with management. Managers regard training as a problem that must be dealt with—quickly and all at once—so that the new trainee can be brought up to full productivity as soon as possible.

Getting employees to do things right means taking the time to train them properly from the start so that they understand what needs to be done, how to do it and why it should be done that way. Effective training, however, involves more than simply providing information. Training is not a problem; it cannot be "solved" and then forgotten. Managers and supervisors at every level must soon realize that training and learning are continual processes that never stop.

The most effective training technique is interactivity. Get people to stand up and do things. Show them how to set a table, look for lipstick on a glass, wash their hands properly, use a thermometer, wash a dish, and garnish a plate. Let the employees participate.

Most managers and supervisors think of training as teaching new employees *skills*, such as dishwashing or cake decorating. Training needs to be far more than that; management must look beyond its own interests. As mentioned before, we must start to consider the employee's interests, goals, needs, and desires if we are to become successful.

The employee must know not only her job and how to perform it, but how her performance affects others in their jobs in other parts of the bakery. She must visualize her position as an integral part of an efficient machine, not as a separate, meaningless function. For example, take the plight of the dishwasher in most bakeries. Dishwashers are vitally important to the success of any bakery, and yet few managers, and virtually no other employees, are consciously aware of their importance. Rather than being treated with dignity and respect, they are considered, in most establishments, insignificant menial laborers. They are often paid minimum wage with little—usually no—benefits, and expected to do all the dirty work; cleaning up after others and working in poor conditions, while all the other employees shout orders and instructions. The only time they are really communicated with is when they do something wrong, or when someone needs something done or a mess needs to be cleaned up. Is there really any wonder why an entirely new crew will have to be trained in two weeks? Many managers themselves don't fully realize the importance of this function, or that it is far harder to find a good dishwasher than it is a good waitperson. You should have every new hired employee perform at least one shift in this position to fully understand its importance. Try giving the dishwashing staff an hour-long break one night and see the resulting chaos.

Telling an employee that his position and performance is crucial to the bakery's success and *showing* him the reasons why are two entirely different things. The importance of performing his job in the manner in which he was trained must be physically demonstrated to the employee, as well as the ramifications of varying from these procedures. Using the example of the frustrated dishwasher, let's apply this philosophy with some practical, hands-on management.

Show them that you are concerned with both them and their performance. Pay more than the other bakeries in the area so that you will attract the best people. Set up some small benefits such as a free meal and free soda per shift. A financial incentive is the most effective type of motivating force. Establish bonuses for the dishwashers, such as giving them a little bit extra for work on particularly busy nights. The small cost of these little extras will be substantiated with lower turnover rates and higher production.

Apply this principle of demonstrating rather than lecturing to illustrate your points with all of your employees, and you will have the basis for a good training program and good employee relations.

Orientation And Instruction

A complete new hire orientation is an intricate part of the training process. The entire orientation will take less than thirty minutes; unfortunately, however, it is rarely done. There is no excuse for not giving the new employee a good introduction before he or she starts the actual training.

Described below are some basic orientation practices:

1. Introduce the new employee to yourself and the company.
2. Introduce the new employee to all of the other employees.
3. Introduce the new employee to his trainer and supervisor.
4. Explain the company's employee and personnel policies. Present her with a copy and have them return the next day to sign the "I have received and read" commitment.
5. Outline the objectives and goals of the training program.
   a. Describe the training, where and how it will take place.
   b. Describe the information that will be learned.
   c. Describe the skills and attitudes that will be developed.
6. Set up a schedule for the employee. It should include:
   a. The date, day and time to report to work during training.
   b. Who will be doing the training and who the supervisor is.
   c. What should be learned and accomplished each day.
   d. The date when the training should be completed.

Ideally, the employee's regular supervisor does all of the training. The trainer must be a model employee who is thoroughly knowledgeable about and experienced in the job. He must be able to communicate clearly and have a great deal of patience and understanding.

The trainee must be taught the *how, why, when* and *where* of the job. This is best accomplished by following the trainer's example and methods. After confidence is built, the employee may attempt to repeat the procedure under the watchful eye of the trainer.

The trainer must gauge how fast the trainee is learning and absorbing the material against the time schedule set up by the manager. Daily written or oral reports should be presented to the manager on the trainee's progress and needs and placed in her employee file. Compare the trainee's production to that of an experienced worker. Written and practical tests can be given to evaluate how much material is being absorbed and utilized. Add new material when the old material has been assimilated. Relate the old material to the new as you continue to build towards training the employee.

Once the employee has completed the training, the trainer or supervisor should prepare a final written report and evaluation. This report should describe the strengths and weaknesses of the trainee, her knowledge of the job, quality of work, attitude, and a general appraisal of the employee. After the manager reviews this report, all three parties should meet to discuss the training period. The employee should be congratulated on successfully completing the training program.

A review of the final report would be in order and then filed in the employee's personnel file for future reference. Ask the employee her opinion on the training program; she may have some thoughts on improving it. This same question should be presented to the employee after two weeks of work. Find out if the training program adequately prepared the trainee for the actual job.

Outside Help In Training

When training sessions involve several people or even the entire staff, you might find it helpful to bring in outside support for your meetings. Look for business associates and experts in their fields who are interested in sharing their wisdom. Often just a phone call is enough inducement. Reward these people for their time and effort with a complimentary gift certificate.

There are other great resources for outside-training information available to assist in your training programs: videos, posters, books, software, etc. One great source for all these products is Atlantic Publishing (**www.atlantic-pub.com** or 800-541-1336).

Below are some ideas for speakers and topics to enhance your staff's knowledge of running the bakery:

- **Health Department Inspector** – Health and sanitation practices/ requirements.

- **Dishwasher chemical company salesperson** – Proper use of machine, water temperature regulating and chemicals.

- **Equipment manufacturer representative** – Proper use of equipment, maintenance and cleaning tips.

- **Community Service Police Officer** – Crime prevention and response tips.

- **Fire Department Instructor** – Fire prevention and emergency response tips.

- **Bank Trainer** – Proper cash handling techniques, counterfeit bill recognition.

- **Ergonomic Advisor** – Assess and review how people interact with machinery and facility. Teach people how to properly align body to minimize physical stress and fatigue.

- **Red Cross Instructor** – Basic first aid, the Heimlich maneuver to stop choking and CPR procedures.

Evaluating Performance

Evaluating each employee's job performance is a crucial element in developing a productive work environment and sound employee relations. Every employee must be aware of his or her strengths and which areas of his or her job performance need improvement.

Quarterly or periodic one-on-one evaluations help break down the communication barriers between management and employees. Many of an employee's work-related problems, thoughts, and ideas can be revealed in the evaluation session. However, remember to keep in mind that evaluations are only a part of the communication process and should not be considered as a substitute for daily communication. You and your supervisory team must always be available to listen. Communication is an ongoing and continual process.

Consider these points before filling out the evaluation forms:

- Know the employee's job description thoroughly. You are evaluating how well the employee meets the job requirements; you are not comparing against other employees nor evaluating the employee according to what you see as his potential.

- Always conduct the evaluation in private with no interruptions. Schedule each evaluation far enough apart so that there is plenty of time to discuss everything in one sitting.

- Don't let just one incident or trait—positive or negative—dominate t he evaluation. Look at the whole picture over the entire time since the last evaluation.

- Evaluations should balance positive and negative attributes. A primarily negative evaluation will almost never motivate a poor employee. Bring out some of her positive contributions and describe in detail what changes are needed. A completely negative evaluation will only scare the employee. Should a very negative evaluation be warranted, it is probable that the employee should have been terminated long ago.

- Review past evaluations, but don't dwell on them. Look for areas where improvement or a decline in performance has taken place.

- Always back up your thoughts and appraisals with specific examples.

- Allow plenty of time for the employee's comments. Remember, you could be wrong.

  o If examples or circumstances that were never mentioned before come out in the evaluation, you are guilty of allowing the communication process to deteriorate.

- Don't cover too much material or expect the employee to make a drastic change overnight. An evaluation is only one part in a series of continuous steps to direct the employee.

- Begin the evaluation with the employee's positive points and then direct the discussion to areas that need improvement.

- Certain personality traits and deficiencies may not always be changeable. Don't overemphasize them, but show how they might affect the employee's job performance and the performance of others.

- Finish the evaluation on a positive note. The employee should leave with a good feeling about his positive contributions to the bakery and know precisely what and how to improve on his weaknesses.

- After the evaluation make certain that you follow up on the thoughts, ideas, and recommendations that were brought out during the evaluation. Without a follow-up, the evaluation is of little value.

- Evaluations are confidential. File them in the employee's personnel file only if no one else has access to them there.

Scheduling

The overall objective in scheduling is to place the most efficient employee at the job and shift where he or she will achieve maximum productivity at minimum expense. The greatest tool management has in controlling labor cost is scheduling, and yet scheduling is most often so poorly done that it becomes more a part of the problem than of the solution. In many cases the employee's schedule is scribbled on a piece of paper or, worse, verbally communicated with little thought as to what is actually needed.

Properly preparing the weekly schedule for a bakery of 15 employees may take between an hour to complete. The individual preparing the schedule has to take into account many different factors, such as:

1. Peak periods (specific hours or days of the week).
2. At what time maximum production must be reached.
3. Special events that need additional assistance or will require staff to leave the premises.
4. The skill and productivity of each employee.
5. Each employee's desired schedule: days off, hours, etc.

Only after several months of operation will you be able to accurately assess your precise labor needs. During the first couple of months be sure you have plenty of employees available should it become suddenly busy. Many customers will understand that you have just opened and don't have all the bugs quite worked out yet, but they are still paying full price for everything. Don't get caught short on trained personnel. Your service on opening day should be consistent with that several months later.

Schedule employees throughout the day to meet anticipated needs. Schedule your best employees to open the bakery, then schedule the other employees in sequence at the time you need them most.

Full-time employees should be cross-trained to perform two or more jobs so absences are less burdensome to the remaining staff and, most importantly, to your customers. There are many beneficial results from this situation. Scheduling will become a lot easier if you have some employees willing to be shifted around to meet your needs. Employees who call in sick or leave without notice can be replaced easily and without disrupting the entire schedule.

Many employees will enjoy performing two jobs, as they will not become as easily bored and will tend to feel like an integral part of the bakery. An employee who is involved, interested, and concerned about the bakery will always be a better performer than one who is detached. The only possible disadvantage to this cross-training arrangement is that, when the employee does leave, he or she will be harder and more expensive to replace. Perhaps this thinking is a great part of the labor problem the food-service industry faces today. If you never give the employee the opportunity to develop and prove herself because you are afraid she'll only leave in a few months, undoubtedly she will. Every employee should be given the opportunity for more and more responsibility.

Should over-scheduling occur, employees can be set at tasks that will produce future labor savings, such as preparing nonperishable food items or cleaning and organizing work areas. Additional unneeded employees can be sent home. Check with your Labor Board regarding these regulations. Under most state rulings, the employee who is scheduled to work and then is not needed must be compensated in some manner; usually he is paid a minimum of three hours' wages.

The bakery's sales history is another important tool in scheduling for productivity. Don't forget special events. For example, large parties, holidays, weddings, and tourist events are as important to anticipate as business downturns.

In recent years, computer software has helped management enormously in scheduling employees. This software simplifies the time-consuming, labor-intensive manual processes involved in scheduling employees, maximizing resources, controlling labor costs, and retaining a qualified staff. Most software systems enable the manager to quickly determine the right employee for each position and shift. Most of the software is easily integrated, with time, attendance, and payroll systems.

All of these systems will save money and enhance employee satisfaction and retention. The software will range in price from a few hundred dollars to several thousand dollars.

Terminating An Employee

There comes a time when an unsatisfactorily performing employee, after being evaluated and given a fair opportunity to correct his or her deficiencies, must be terminated. Discharging an employee is always a difficult and unpleasant task, but it must be done for the good of the bakery. Although it may be an unpleasant experience, it is far worse to let the employee stay on. Before long, the entire staff's morale will drop, causing a decrease in productivity.

The decision as to whether an employee should be terminated or retrained is difficult and often prejudiced by your inability to examine the entire picture of the employee's performance. The final decision to discharge an employee should be reached after carefully weighing the pros and cons—never in anger or when tired or under stress. Ask the employee's supervisor for an evaluation of the situation and the employee. Examine the employee's training, supervision, and past evaluations. Make certain the employee has been given a fair opportunity to prove himself. Also, be certain that neither you nor any member of your management staff has in any way contributed to, caused or perpetuated the problem.

Immediately after reaching your decision to terminate the employee, set up a meeting with him. Don't let more than twenty-four hours go by; you don't want this information to leak out.

Exit Interviews

An exit interview is a valuable tool for the employer and the employee. You may learn things that you were previously unaware of and can correct with future employees and they may learn how to become better employees in the future.

- The employee's supervisor should be present during the exit interview. She will be able to add support and witness the action. This is important, as the employee may use some legal means to gain a settlement.

- Conduct the exit interview in a private room with no interruptions.

- Should the employee disagree with your reasoning or points, give him the opportunity to discuss them, but make sure you back up everything you say with proven facts and statements.

- Remain seated and calm during the proceedings; don't get up quickly or move suddenly.

- Never touch the employee, except when shaking hands. These actions may be misinterpreted and lead to a confrontation.

- Fill out a report on the termination proceedings, and file it in the employee's personnel file. This report will be important should the employee decide to challenge the action.

- Develop a plan to fill the vacancy as soon as possible. Keep in mind that it will take several months before a new employee can be brought up to full productivity and that, after training, he or she may not work out at all.

Although nothing can fully prevent a former employee from filing a lawsuit, there are ways to decrease the likelihood of litigation: Be honest with the employee about his or her performance and the reason for the termination; treat employees consistently; and aggressively investigate claims of discrimination.

Above all, document the chances you've given employees to improve prior to your decision to terminate them. If a termination is challenged, and there are no records of the problems cited as reasons for termination and indicating opportunities given to correct these problems, there could be a problem. Keep in mind your notes don't have to be very detailed, but a dated description of a problem or of the employee's progress that is slipped into a personnel file helps dramatically.

Your other employees always perceive the termination of an employee as a threat to their security. You may even be looked upon as unfair or exceedingly harsh. Sometimes an explanation is needed to soothe the other employees. Never share derogatory thoughts, discriminatory comments or confidential information to your other employees. In most cases, though, the reasons will be obvious to them, and they will be on your side.

Still, *document everything.*

Chapter 7

Your Customers

s a culinary artisan, you devote hours to creating the perfect sour-dough loaf, the freshest pies and nutritious whole wheat breads, but unless you are always creating happy customers, your business will fail.

As a new bakery, you'll need to entice more than family and friends to your establishment to keep the doors open. As a new owner of an existing bakery, you'll need to keep current customers happy (some people hate change) and to find new ones interested in your product mix.

Who are Your Customers

In researching and writing your business plan, you'll have learned tidbits about who your potential customers are. Now may be a good time to create an "ideal" customer. Understanding this person will help you create the products they want, at the highest price they will pay, with the service they expect.

As a new business, you will probably start with some assumptions that you'll need to assess and fine-tune after your doors are open.

- Think about the type of people who are coming in.
  - o Look at how they are dressed - upscale singles or casual families?
  - o Do they want individual desserts or jumbo rolls?
  - o How long do they take to make up their mind?

- Why they are coming in?
  - o Buying for the family or office?
  - o Need something for a picnic or holiday?

- Where they are coming from or going to?
  - o On their way home from work? After church? After a movie?
  - o How they arrive (walk, drive, bus) – special packaging needed?

- How did they find you?
  - o Are they coming to you through a wedding coordinator?
  - o Do they bring in your newspaper coupons?

If you don't know much about your customers – ask! Introduce yourself and ask questions like:

- Is this your first time in?
- Do you live nearby?
- How was your food/drink today?

People, People, People

Presence

The food-service business is about personal connection, and, of course, it's about food, too. If you can establish a connection with your customers, you will exist in their minds as a welcoming friend of the family, not just a place to buy a loaf of bread. People want to be treated like individuals, and they will repeatedly do business in a place that treats them well.

So how do you connect? By giving your full attention to your customers; multi-tasking is not the way to go. If you're taking a phone order, putting tonight's specials on the board, and talking to your new cookie baker, how much real attention will any of those tasks get? Do you think the potential customer on the phone or your employee will feel well treated? Will you have time to check the spelling on the specials board?

The truth is that we can really only focus on one thing at a time. So when you talk to your staff or your customers, *really* talk to them: *Listen,* say what needs to be said, and move on. It's the same with the specials board. Drop distractions, handle each item individually and then move on to the next. Presence is simply a lack of distraction. If you act distracted with your staff, they will keep asking the same questions or coming to you with the same problems. If you are distracted around your customers, they won't come back at all. Pay attention. People will notice, and you will too. Be a great listener.

Getting To Know Your Guests

In a business that only thrives through personal connection, getting to know your guests is crucial. Go beyond the procedures of service, and start thinking of your customers as individuals. There is a difference between serving 200 pies and creating Steve and Mary Carson's 50th wedding anniversary cake or sponsoring a fundraiser for the Friends of the Museum of Science.

Numbers are important, but your relationship to your customers drives your business. Furthermore, the two easiest things to learn about your customers are also the most useful: *who they are* and *what they enjoy*.

People love it when you remember who they are. It instantly makes them feel like they're insiders and makes them feel important in the eyes of their friends. Remember Norm from *Cheers?* Norm felt pretty comfortable there, and he definitely went back to the bar very often. As a manager, you probably know your regulars by name, but do you have a system in place that teaches your new staff who these important folks are? If you were a regular customer at a bakery and an employee you had never seen came up and greeted you by name, wouldn't you feel special?

You can train your employees to ask people's names and take time to remember faces. Have your customer service staff wear name tags – this is a great way to establish a rapport. Continually remind wait staff that they are serving people, people, people— not anonymous mouths. Using guests' names is another win-win proposition, because the more you use their names, the easier they will be to remember and the easier it will be to treat them as individuals…which brings you to the next step.

Now that you are talking to customers as individuals, the next step is to find out what they want as individuals. How? You must ask, but you also must remember, not only what you've been told, but also what you've observed. If you are doing a good job (good food and good service), you'll soon have regulars. Remembering that they prefer decaf coffee or no onions on their sandwich transforms anonymous faces into valuable long-term customers.

Appreciation

What do you do to let your customers know that you appreciate them? If you recognize them and make them feel important, it will draw them closer to your bakery and further differentiate you in their eyes. The following list touches on a few tried-and-true examples. Many can be done at little or no cost. However, unless you've actually taken time to get to know them – you won't be able to make your appreciation personal.

- Put them in your newsletter (You can have either a print or an e-mail version).

- Put them on a Wall of Fame or "Outstanding Customer" plaque.

- Give your regulars awards, and/or honor people in your community who make a difference through charitable work.

- Name menu items after guests. Customers LOVE this, and who knows what soon-to-be famous culinary delights are cooking in their minds?

- Personalize booths or seats.

- Put guests' names up on your Bulletin Boards.

Clubs

Clubs can be a great way of treating your regulars as individuals and giving them privileges or paraphernalia other customers don't have. If you sell coffee (dine in or take out), here are a few ideas for having a great and effective "Mug Club":

- Keep mugs or cups on display to promote your Mug Club.

- Have the mugs be so distinctive that other customers ask about them and the people with the mugs feel special.

- Put customers' names on their mugs.

- Give them a deal: cheaper product, more for the same price or another incentive.

Do you sell donuts, bagels or fabulous milkshakes? Do you sell anything else that could be a club people would join and, through it, accrue points towards a prize? They could get a discount on baked goods and a free gift after they've eaten, say, 50 donuts. Clubs can be a great way to distinguish your products, distinguish your customers, and give them a sense of belonging. The results will benefit your bottom line.

Bringing in New Customers

Your advertising and public relations efforts along with building a great word of mouth is how you'll bring new customers to your bakery. In Chapter 8 and 9, you'll find marketing and public relations ideas and resources. One of the best ways to bring in new customers is to get your current customers talking. Word-of-mouth advertising is invaluable. Someone they trust – a friend, a neighbor, a coworker or a family member, relays the information. Word-of-mouth advertising will be discussed later in this chapter.

Keeping your customers happy is a great way to create a positive "buzz" about your products and service. The remainder of the chapter will talk about good communications and delighting customers. Do that and you'll have all the buzz you'll need!

How to Keep Customers Coming Back

Growing your business requires an ongoing effort to bring new customers to your door. It can take a significant investment – in time and money – to keep your name in front of *potential* customers. No matter how many new customers you bring in every day –**how you keep them coming back will determine how profitable your business will be**.

Investing your resources in building long-term customers is significantly less expensive (and more practical) than advertising for new customers. Even the largest city has a finite number of potential customers for your business so once they grace your doors – you've got to keep them coming back.

You cannot succeed as a stand-alone bakery without excellent service. Variations of your entire product line are readily available at the nearest supermarket. Although your handcrafted desserts and baked goods are superior to those mass-produced goods, it's also your customer service that will keep people driving by other stores to pick up a loaf of French bread or a package of rolls.

Profitability is what's going to keep you in business, obviously. Are you creating an environment that leaves patrons feeling served and eager to come back? Are you holding staff meetings that leave your crew energized or deflated? Are you, your management and your staff working independently or as a complete whole towards a common goal? This chapter will discuss ways you and your entire staff can create an environment that will keep customers coming back.

Offer Free Wi-Fi

Offering free Wi-Fi access for customers is a great way to build repeating business. Whether they are leisurely enjoying the web or conducting business, having the Internet will attract people to your café and make them stick around after finishing their food or drinks.

Communicating With Your Guests

First impressions mean everything. If your customers have to battle heavy traffic or arrive after a hard day's work, your first impression on them is even that much more important. It's an opportunity to let them know their presence is welcome and your luscious desserts and fresh bread are joyous diversions from a tough day at work.

How you relate to your guests affects their opinion of you. That opinion then translates into potential loyalty, and loyalty boosts your bottom line. In fact, a 5-percent improvement in customer retention translates into a 15–50 percent boost in profits. Those are serious numbers. In common terms, that simply means getting your regular customers to return one more time per month. Furthermore, it costs about five times as much to attract a new customer as it does to retain an existing one. This is another huge benefit of loyalty to your bottom line, and it comes through the overall commitment your establishment makes to its repeat customers. Focusing on your repeat customers – your most profitable customers – allows you to keep them coming back.

Two things to focus on for retaining customers:

1. Pay attention to your most profitable clients. Listen. Keep in touch. Find out what they want and need and why they've chosen you.

2. If they go to your competition, find out why.

So how do you gather this information? In an establishment where people come and go quickly, this can be a challenge. Below are a few ideas to help you learn more:

- Tuck a prepaid postcard survey into their bill. Ask your customer to rate your bakery, your food and your staff. The one question that must be there is "Would you visit us again?" If you get a "no," take immediate action to determine why and then fix the situation.

- Provide a comment section on your website.

- Create comment cards where customers rate your bakery. Have your staff hand these out to people over the counter or include with delivery orders. Offer discounts or promotional items for the return of the cards.

- Make follow-up calls to wedding planners, brides, caterers and others who order special customized cakes or desserts.

- Create a customer database to develop a line of communication through newsletters, remembrance programs (free cupcake on your birthday) and mailings.

When communicating with your customers always:

1. Thank your customers for their business.

2. Whenever you can, individualize your communications.

3. Make it easy for them to communicate with you. If it is only one way, it isn't truly communication. If their opinions are important to you, they'll come back.
   a. Include your phone number and Web address on all packaging, promotional materials, and brochures.

4. Listen to and act on customer suggestions.

5. Inform customers on new or improved services. Put your PR wheels in motion.

6. Tell customers of potential inconveniences like renovations, and stress their future benefits.

7. Answer every inquiry, including complaints.

8. Accommodate all reasonable requests for substitutions or special orders.

9. Empower employees to solve problems.

10. Talk to your customers and employees so you can let them know you're listening and find out what's going on.

Customers For Life

Take care of your guests and your sales will take care of themselves. "Customers for life" means that once guests come to your bakery, they'll never be satisfied with your competitors. Simple, right? It also means that the real work of building sales doesn't happen with your advertising schedule or marketing plan, but face to face with your customers.

The key to building bakery sales is to increase volume from your existing customer base. Think about it: if your customers were to return just one more time per month that would be an increase in sales volume of between 15 and 50 percent! These are people who already know about you, live within an acceptable travel distance and will recommend you to their friends if you make them happy.

So how do you do this? Work on building loyalty, not just your per sale average. It's true – a bigger check is a bigger sale, but your income comes from serving people, not food. Focusing only on the bottom line puts your customers second at best. If everyone who ever ate at your bakery were so pleased that he couldn't wait to come back, what would your sales be like? If buying from you isn't a pleasant (and perhaps even a joyful) experience or if they felt pressured to order something expensive, what difference does it make how big their order is if they won't be coming back?

This isn't to say that suggestive selling can't work. If it's done well, it can be very effective. However, it can backfire if it appears the goal is to just sell more, more, more. Your goal is to delight them, win their loyalty and put them first!

Incentives

Incentives work because people do what they are rewarded for doing. It is as simple as that. Reward customers for coming back and they will.

There are three basic ways to do this:

1. Discounts.
2. Promotions.
3. Customer loyalty programs.

Discounts

An effective discount actually *gives your guests a deal* and *generates more profit for you*. How? By making a sale you wouldn't have made otherwise. Is a customer buying a discounted cake that, even with the discount, has a 40 percent profit margin, instead of the cookies that she would have ordered and that has a 30 percent profit margin? Then she's getting a deal and you're making more money. Are your business card drawings giving customers a chance to buy items at half price, but bringing in more than twice the business or bringing people in during non-peak hours.

Internal Coupons can be a great way to increase repeat business. Three of the most widely used are:

1. **Courtesy coupons.** These are wallet-sized coupons that your staff carries. They can be issued to customers and used on return visits. They are great if a customer has a complaint or is put out somehow, or they can be used to reward customers for their ongoing patronage.

2. **Cross-marketing coupons.** If you have very fast and very slow service periods, why not offer a discount to customers if they return during the slower time? If you do a great breakfast rush, for instance, give people a free afternoon snack if they come in on Wednesday afternoons.

3. **Companion coupons.** Encourage your regulars to bring a friend. You can offer a two-for-one special on lattes or milkshakes.

Promotions

Five great promotional opportunities are:

1. Birthdays.
2. Anniversaries.
3. Holidays.
4. Special Events.
5. Festivals.

Birthdays and Anniversaries: Do you have an irresistible offer for patrons who celebrate their birthdays or anniversaries with your cakes? You can get the dates of their special occasions when they sign up for your Frequent Buyer program. (Hint: Let them know when they sign up that you're asking about these dates in order to offer them specials on their birthdays and anniversaries; otherwise they could feel their privacy is being invaded.) With this information you can invite them to celebrate with one of your creations. Also, make sure your offer is valid for more than just the actual date being celebrated—within the month is effective—because people need some flexibility in planning their special events.

Holidays: The beauty of the holidays is that someone else advertises them. You don't have to tell your customers Thanksgiving is coming, but you could put a flyer in with their October orders about the pies you'll have on special or instructions on pre-ordering.

Special Events: Special events can be a great way to promote business and good will. Have you thought of hosting a tea tasting, an all chocolate celebration, participating in a community festival or a charity fundraiser?

Special events can be held during after-hours (especially for bakeries that close early in the day) where attendees can freely mingle. Theme events are a great way to bring people in that might never have stepped through your door. These events can be held merely for the fun of it or you can use the occasion to recognize employees or community leaders, or as fundraisers.

Product festivals are great reasons to invite customers to come back, and they're great things for customers to talk about. They highlight specific cuisines or products and can be a great way to stir things up for staff and customers alike. You can run a festival on a specific night or for a specific time period—usually a week or two. Make sure however, that the festival you hold is right for your establishment, and that it is run frequently enough to break up your routine, but infrequently enough to remain special. Do a memorable job and build a strong foundation for future events.

Product festivals usually coincide with seasonal items—raspberries, pumpkins, etc.—when the items are abundant and cheaper. Off-season festivals can be great for word of mouth—if you can find the product. People would love a fresh strawberry festival in January. Regional food items like King Cakes for Mardi Gras and Navajo fry bread might be just the ticket to attract new customers.

Community celebrations, festivals, farmers markets, and fairs can be a good venue for you to introduce your products to others beyond your neighborhood. Thoroughly research these venues before committing your staff to these labor-intensive events. Whether you are bring previously baked goods or cooking on-site, you must comply with all state and local food service and health regulations. Stay away from sponsoring non-food booths as your goal is to get people to *sample* your baked goods and come into your store for more!

A *charity fundraiser* can be a way to gain exposure for your bakery that pays off for everyone. It will improve your image and distinguish you from competitors. It doesn't have to be a break-even venture either, because you only give a portion of the proceeds to the cause. Ask your suppliers to donate supplies or share the duties in exchange for promoting their wares and community good will. Enlist neighboring businesses to lower your costs and increase the donations. By working with complementary businesses, you also increase your chances of getting some in-depth news coverage of your event. The charity will also promote you to its supporters, which can bring new people through your door who want to support your business.

For more information on charity and PR, see Chapter 9.

Frequent Buyer Programs

Many businesses have found frequent-buyer programs are an excellent way to build loyalty and repeat business. Your frequent buyer program should be simple for you and your customer. If you can get people to "sign up" for your program, you'll have a way to keep in touch. Ask for basic contact info (including email if you wish to email specials or a newsletter) and special events (birthdays, anniversaries) that warrant a special remembrance or discount.

Rewarding your customers for continued loyalty gives them an added incentive to choose you over the competition and will help bring them back that extra time per month. They usually come in variations on three basic forms:

1. **Punch cards.** An inexpensive card that is typically issued for free and is punched every time the guest purchases a product. When they have purchased a certain number of items they receive something for free. The biggest plus of punch cards is their ease to produce. The biggest negative is the ease with which they can be altered. Keeping the guests' cards, or duplicate cards, on premises can help counter this.

2. **Point systems.** These are often dollar-for-point systems, in which a customer accrues points towards free food or merchandise. This can be a great way for guests to "eat their way" towards a free gift or party desserts. A point system is considerably more complicated to implement than a punch card system, often including outside vendors.

3. **Percentage-of-purchase programs.** This is the closest type to the airline programs, with guests paying full price for items while accruing dollar credits for future purchases. This gets people in the habit of thinking of their purchases as having a larger than normal value and keeps them coming to you. This type of loyalty program is probably best suited for wholesale buyers (corporate dining rooms, local restaurants) of your goods.

Punch cards, point systems, and percentage-of-purchase programs are all ways to monitor your guest's patronage, reward them for coming back, and increase your opportunities to delight them with your food and service. Take some time to figure out which is right for you.

Delighting Your Guests

Expectations

Satisfaction isn't even close to good enough. It's an improvement on dissatisfaction, of course, but in today's market, it won't keep people coming back. There is just too much competition. You need to *exceed* your guests' expectations, every time. The food-service business is built on personal connections. You serve one person at a time, and the more personal that interaction, the more you'll exceed her expectations—and the happier she'll be.

Here is a list of basic customer expectations and some hints on how to meet and/or exceed them. Your customers expect:

- Hot foods to be hot, and cold foods to be cold.
- Fresh coffee. Anticipate their needs and have a fresh pot ready.
- First come – first serve. It can be difficult during peak periods to keep track of who entered first but don't let pushy customers or your inattention make someone wait any longer than absolutely necessary.
- To pay quickly. Be certain that your cash register is properly stocked to give change and manned by a well-trained person with cash handling experience.
- To know how long it will take to prepare, package or customize something while they wait.
- Your staff to know the menu and how baked goods are prepared, and to be able to answer questions about ingredients.

- Your staff to care more about them than when their next break is or what they saw at the movies last night.

Do you know what your customers expect when they come through the door? Are you out to exceed those expectations and give each guest a memorable and delightful meal every time?

Ways to Delight

Customers are delighted when you care – it's as simple as that. Doing things that demonstrate how much you care will make a difference. Part of the trick here, however, is that there is no trick. You've got to be sincere. People know when they're being treated with sincerity or with a mechanical technique. Sincerity works. Here is a list of practices that, when done with sincerity, can give guests a feeling of being taken care of, given real value or simply delighted. These touches may appear to guests to be extraordinary or creative—things that they never would have thought of themselves:

- **Provide umbrellas when it rains.** Keep an extra-large umbrella handy to protect your customer while she carries her dessert to her car.

- **Free stuff while they wait.** People mind a wait a lot less when they've got a complimentary glass of warm cider to keep them toasty and a cookie or the local paper or a magazine to read. They will appreciate you going the extra mile.

- **A place to rest and wait.** If your bakery does not have a dining room, provide a couple of chairs for waiting. A bench or "ice cream" table outdoors also makes a nice place for people to enjoy a one of your hot-from-the-oven treats.

- **Connect them to the Internet for free.** Almost every customer relies on the Internet to conduct business or stay connected with people. Offering free Wi-Fi is a great incentive to attract customers to your café.

- **Introduce yourself.** People like to meet the person in charge. They appreciate that someone important is checking in on them.

- **Share "secrets" with them.** Got some new cookie debuting next week? Why not give away free samples today to whet people's appetites? Don't forget to ask their opinion and give them a discount coupon for next week.

- **Offer reading material.** Have a "lending" library with books, magazines, and newspapers.

- **Free postcards and postage.** Do a lot of tourist business? Why not give them stamped postcards (depicting your bakery, of course) for sending their "Wish you were here" messages? It's a very low price to pay for something they'll appreciate and advertises your bakery all over the world.

- **Provide map directions on your website.** Have a great, clear map on hand, and when callers ask for directions to your bakery, point them to your website or give them directions on the phone.

- **Provide armchairs for the elderly.** It's harder for the elderly to get in and out of their chairs. If you serve a lot of elderly customers, or even a few, have chairs with arms to make it easier for them to get in and out. Let them know you did it just for them. They will certainly appreciate it.

- **Display a guest book.** Make sure your guests fill in the guest book: you need a mailing list of your patrons for sending them promotional material. Try to collect birth dates and anniversaries for your database, as well.

Word-of-Mouth Advertising

Positive word-of-mouth talk is the best advertising there is, without question. Does it just come by accident or only from serving great food? Yes and no. Great word of mouth comes from your customers having something great to talk about and their sharing it effectively. Do you have a deliberate, creative, and authentic plan in place to create great word of mouth? You can and should have everything to do with whether your guests have something to say and whether or not they're saying it.

Customers don't talk about you unless they're thinking about you. You want them thinking about you in the right way, which means you have to educate your customers on why they come to you. To do this, you must create points of difference between you and your competitors. Then people can tell their friends about why they order a cake or drop by your bakery for a five-grain bread.

An effective word-of-mouth program has five main goals:

1. Inform and educate your patrons.
2. Make the customer a salesperson for your bakery.
3. Give customers reasons to return.

4. Make your service unique and personal.
5. Distinguish your business from the competition.

Points Of Difference

If you want your customers to return one extra time a month and tell their friends and family about you, you first need to distinguish yourself from the competition. You do this by creating "points of difference."

What is different about your establishment? Your concept? Your freshness? Your artistic skills? Do you guarantee your service? Give free coffee to waiting customers? Have several organic choices? What makes your place memorable and different from the competition?

Here is a partial list of things that you can develop into points of difference from your competition.

*   **Water.** Serve local spring water or imported bottled water, or simply filter your tap water so it tastes good. Put a lemon slice or food-safe flower petal in the glass or carafe.

*   **Soft Drinks.** Serve bottled drinks instead of post mix. Have an extensive selection and offer free refills.

*   **Special ingredients.** Have unusual, local and/or organic ingredients. Promote healthy choices and allergy-safe alternatives.

*   **Restrooms.** Have twice as many restrooms for women. It tends to take them longer; why should they have to wait?

*   **Delivery service.** Offer deliveries for large party orders, oversized items (cakes), office parties, and special events.

*   **Birthday candles and matches.** Include a free box of birthday candles and matches (these can be imprinted with your bakery logo and address). Fewer people smoke nowadays so matches aren't always handy.

Certainly not all of these are appropriate for every bakery, but finding a great way to distinguish yourself – often through a mishap or brilliantly ridiculous staff idea - is a great way to give your place a real identity and give your customers something to talk about.

Educating Guests On The Differences

Having a great idea in place isn't enough, though. You've got to inform your customers about it and give them the words they can then pass on. A customer telling his friends he had a great chocolate cake is great. A customer telling his friends he had an elegant cake with fresh local raspberry filling and imported Swiss chocolate is worth his weight in gold. Details differentiate your product and make yours the place to go for something *extraordinary*.

How do you get this information across? Arm your staff with words they can comfortably work into a conversation. Do you offer milk-free products? When guests call and ask say, "Dave's Bakery offers loaves and rolls that are milk-free. We're the only place in town that does that."

An effective word-of-mouth program not only creates points of difference between you and the competition; it educates your guests on those distinctions. If you give your customers a great experience, and the words to describe it, they'll talk about it to their friends. The first step is to educate your employees on these points. A good way to put together a list of these differences is during a staff meeting when ideas can be tossed about and descriptions created.

Your Staff Makes the Difference

Your bakery is made up of two things in the eyes of your customers: the food and the staff. The quality of service your customers receive will determine their opinion of your bakery. Your employees are the ones who delight your guests—or don't—who give them things to talk about and who provide the crucial personal connection. Staff will execute most of your sales promotions and programs, educate your customers about what makes your bakery better than the one down the street, and give your guests information they can pass on to their friends.

It's in your waitstaff's best interest to connect with customers. But—here's the thing— *your staff will treat your guests the same way you treat your staff.* If you want your staff to be gracious, to listen, and to delight your guests, you have to do the same for them.

- **Greet guests within a minute.** Don't leave them waiting. Waiting will negatively affect a customer's mood. Your bakery should be filled with warm thoughts and happy memories.

- **Make eye contact.** Don't stare at the door, the floor or the artwork on the wall. Clear your head, smile and pay attention. Make sure you're *fully tuned* *in* when you're talking. Don't talk to your customers as you're flying by. It makes people feel unimportant, and no one likes that feeling.

- **Focus your energy on taking care of your customers. Try** making them happy, doing little things that exceed their expectations and generally making their shopping/dining experience as enjoyable as possible.

- **Encourage your customer's food choices.** People can be strange about making decisions. The simple act on your part of telling them that you've had what they're ordering and it's great can take away any anxiety they have about making a bad choice.

- **Tell the bakery staff good news.** Just like you need to be sensitive to the mood of your customers, be sensitive to the mood of the bakery crew. The bakers don't want to hear about things just when they're wrong. Pass along good news to them, and they will probably make it easier for you to take great care of your customers.

- **Ask before refilling coffee.** Coffee drinkers can be very particular about the amount of cream and sugar they have in their coffee. Temperature also matters. Don't top off a cup they may have spent considerable time getting just right.

- **Tell guests about specific events at your bakery and invite them.** This is a more effective way to let them know you would love for them to come back and to build a personal connection. It can be much more effective to invite guests to return for your chocolate cake special on Tuesdays than just to say "Thanks. Come again."

- **Show gratitude.** People are dealing with a lot in their lives and you have a chance to "make their day." Express gratitude in the tone of your voice when you thank them for their patronage or invite them to come back. Making them feel appreciated will make them remember you and the bakery.

- **Make personal recommendations.** Tell your customers what *you* like. This is not suggestive selling, because it's sincere and therefore won't alienate your guests. Your enthusiasm will be infectious, even if customers don't order what you recommend. It won't bother them that you're excited about what's on display.

How, as a manager, can you make it easiest for your counter staff and waitstaff to do these things? For one, have them taste everything on the menu. Ideally, they should know how every item is made so they can speak knowledgably about it. Even better, as part of their training, they could work in the preparation area for a day or two. If your staff has together sampled all your food and beverage—maybe you could throw a tasting party where everyone gets to know each other and gets an education—then they will be able to make educated and sincere recommendations. Nothing is more persuasive than a server who knows what he is talking about.

Also, let them use their own words to convey their enthusiasm. It's hard to make a personal recommendation using someone else's words. You want them sharing *their* enthusiasm, not a canned version of yours. Your crew will find their own way of expressing their enthusiasm. Letting them in on what you sell is the best way to give them something to be enthusiastic about.

Motivating Your Staff

How are you going to impart all this newfound wisdom and good spirit to your staff, and how are you going to get them excited about delighting your customers? You need an effective, uplifting staff meeting.

Most staff meetings are far from invigorating. In fact, they usually create an energy dip and a staff that feels like they are on management's bad side. An uplifting staff meeting is not just a gathering of bodies with one person giving out information; it generates a positive feeling within the entire group. An effective staff meeting has three main goals:

1. Generating positive group feeling.
2. Starting a dialogue.
3. Training.

Positive Group Feeling

This will help your staff discover what it has in common and think in terms of working together, as opposed to strictly as individuals. Share good news in order to build good feeling. Staff meetings are not a good time to address individual or group shortcomings. Find the positive—even if you need to hunt for it—and talk about it. This is how you will build a supportive feeling and get people talking.

Dialogue

A good dialogue is a comfortable back-and-forth of ideas that gets people connected and leaves your staff feeling that they're a truly creative part of your bakery. You learn from the staff and they learn from you. Promoting this flow of ideas reduces the "Us vs. Them" mentality of your staff and reinforces team spirit. If everybody is on the same team, service improves while productivity and profits go up.

Training

Good staff meetings are places to share ideas for better performance. This is your chance to pass along tips to your staff while having them learn from each other. Your staff are intelligent people and they instinctively know what works. Encouraging them to share thoughts about work will turn staff meetings into a forum for discussing ideas. This atmosphere will improve their learning curve dramatically.

Ideally, you should hold a quick staff meeting before every shift every day to explain daily specials, discuss upcoming events and put everyone in a positive mood. Longer meetings should be scheduled "after hours" with ample time for discussions and learning (or reinforcing) important or complex topics (safety or sanitation issues for example). Never sacrifice "actual" customer service to hold a meeting on improving customer service! Customers are priority #1.

If you frequently cancel staff meetings, it sends the message that they are not important and that the staff's opinions are equally unimportant. An effective pre-shift meeting should last no longer than 15 minutes. If it's longer, you may lose people's attention—shorter, you won't get enough said. Pick a length and start and finish on time. Include the entire staff. This may be a good time to let servers taste today's specials and have the kitchen staff tell the waitstaff about them.

Possible Format for a 10- to 15-minute pre-shift meeting:

Before you start, remember that the thing that most determines how your meeting will go is your own state of mind. Are you looking at your staff as a group of dedicated people committed to doing a great job or a bunch of goof-offs? Are you a coach on the playing field seeking to facilitate and encourage people's best performances or a judge looking to identify and punish people's mistakes? Rest assured that however you approach the meeting, your staff will feel your mood and it will affect the work they do. Get committed to building on people's strengths and holding energizing staff meetings.

- **Good news (1–2 minutes).** Acknowledge what works and create a good mood. Find something about the business that shows people doing a good job and making guests happy. Acknowledge the doer or bearer of the news with sincerity.

- **Daily news (2–3 minutes).** Outline today's specials and upcoming events.

- **Ask your staff (5 minutes).** This is the most important part of the meeting. This is your opportunity to find out what's *really* going on in your bakery and what people are thinking about. *Listen. Don't interrupt* with your own thoughts and *don't judge* people's comments.
   o   Create a safe space for people to sincerely share what's on their minds and to learn from each other.
   o   How well you listen directly affects how much they're willing to say. Since they *are your bakery*, as well as your access to the nitty-gritty, get them talking. If they're shy, ask them questions.
      - What's working for you guys?
      - What's making things tough?
      - Where have things broken down?
      - What questions from customers have you been unable to answer?
   o   Once you get the ball rolling, you may find it hard to stop! Good. That means people have things to say and you'll benefit.
   o   Asking the rest of the staff if they feel the same way as the speaker is a great way to see if there is a group sentiment and to gauge the size of the issue being presented.

- **Training (3–5 minutes).** If staff comments run over, let it cut into this time. It's important that your staff learns from you, but it's more important for you to learn from them. Plus they will be more open to learning from you, if they know you're listening to them.
   o   Use this time to talk about a single point you want your staff to focus on during this shift, to give out specific knowledge about a product or to train in another targeted way. Focus is important. If you tell people how long the meeting will last and hold to that, they will give you their attention. If you go over, you'll lose their attention and their trust. Get to the point and trust that they got it.

Learning how to conduct productive staff meetings is worth your effort. Developing a good rapport with your employees will make everyone more successful – and your customers happier.

Focus on Making Your Guests Happy

Running a successful retail bakery is about personal connections. That's what sets you apart from the local supermarket. Becoming connected is the way to delight guests and bring them back. Invest in making people happy and you will be rewarded. Remember bringing customers back just *one more time per month* can give you a **15 to 50 percent increase in sales volume**.

If you dedicate your energies towards building an establishment where your staff are treated with respect and gratitude, they will treat you and your customers in the same way. Focus on building an environment that is friendly, helpful, informed and welcoming, and people will come back again and again. This can happen by taking the weight of sales *off* your staff's shoulders. Everybody—especially customers—should feel they are on the same page. People will give if they are given *to* and taken care of, and they will never come back if they feel taken advantage of.

Your job is to create a place that people think of first when deciding where to buy a birthday cake or tantalizing pastries – and that they tell their friends about. Again:

1. Build customer loyalty.
2. Dedicate your business to delighting your guests.
3. Give your guests something to tell their friends about.
4. Give customers incentives to return.
5. Become connected. Your staff is your bakery. Connect with your staff to help them connect with your guests.

Chapter 8

Marketing Your Business

B efore opening your shop, it's important to market your products and services beforehand. You may think to yourself, *I know I need to market my business, but how can I with such a limited budget?* The answer is easier than what you think. This chapter breaks down two types of marketing: Internet marketing and traditional marketing.

Internet Marketing and You

How Can Your Business Benefit from Internet Marketing?

The Internet can help you reach existing customers more easily, and specifically target people you'd like to become new customers. With a little bit of work and guidance, you'll quickly start sharing photos, posting, and building online relationships with customers.

Essentially, the Internet is a personal gateway from a business owner to the customer. Businesses can now understand and cater to people's wants and

needs better because websites gather personal information about every user's likes, dislikes and purchases. Search engines use cookies to record a user's search requests and create ads with the results. Users of social media sites share personal preferences and information freely when they create and use a site. According to a 2014 *Business Insider* article, Americans spend more time on social media than any other major Internet activity, including email. Over 50 percent of people use some form of social media more than once a day.

What is Internet Marketing?

In short, Internet marketing, also known as online marketing, is a tool, strategy, or method that spreads awareness of a company's name on the web. From subtle messages to social media platforms, Internet marketing strategies vary. First, define what your business needs to determine which types of social media platforms and other web marketing tools will benefit your business. According to a recent study by Search Engine Land, 85 percent of all consumers use the Internet to find local businesses. You could be part of that 85 percent if you market your business on the Internet.

There are **three different types** of Internet marketing outlets to choose from:

1) Social media marketing

2) Email marketing and

3) Web marketing

If Internet marketing is a new concept to you, then start with one marketing tool. For example, you can market your company through **email lists**.

As you try different marketing outlets, you may find that web advertisements and emails are more effective than social media or that social media platforms alone work best for your business. Just as every business is unique, so are your Internet marketing tools. What works for some places may not work for yours. Try to be open to new marketing tactics.

Additional Advice: Before diving into more advanced marketing strategies, create a website so customers can see your products and have all the information they need right at their fingertips. Most social media platforms require a link to your website to verify your business anyway, so take some time to get your website up and running. Some **free website builders** and **blogs** include Wix, Sitey, Tumblr, Weebly, Wordpress, and Blogger.

Social Media: Why You and Your Business Need It

Social media are websites and applications that enable people to create and share content or to participate in social networking. Social media tools come in various forms, such as text, status updates, photos, audio, and video form. Today's consumers enjoy reading and viewing visually stimulating information on a constant, daily basis, so that's why businesses promote and sell their products through social media outlets.

According to a 2013 Pew Research poll, about 72 percent of all Internet users spend time on social media every day. Not only is using a social media platform one of the cheapest and fastest ways to market your business, but it also produces the most effective results. If your targeted audience is between ages 18-30, then social media is definitely the best option for promoting your business.

Social media presence allows **local businesses** to advertise to **specific audiences** for *free*. This is the main reason why so many businesses—large and small—are shifting toward advertising on social media platforms. Your potential customers can access your business' information anytime and anywhere; it's advertising made simple and personal for each customer.

Choosing which social media platform to use may seem like a difficult task. If you spend time learning about what each platform offers before making and managing an account, then you can save yourself valuable time and market your business quickly and effectively. The Quick Chart below briefly outlines the Top 10 social media sites.

Additional Advice: Social media is only effective for your business if you use and update it at least a few times a week. Daily updates are best. Review the chart below with the top social media sites and start with one or two. Once you are used to maintaining a social presence you can expand to other sites.

The Top 10 Social Media Platforms Quick Chart

1. Facebook	https://www.facebook.com/business	**What Facebook does:** Reaches a specific, targeted audience for free or small fee for boosted posts.
		How it works: Create a page, post upcoming events, update information, interact with customers, and show off new products.

2. Twitter	https://business.twitter.com/	**What Twitter does:** Reaches out to everyone. Sends out relevant, intriguing information fast.
		How it works: Post tweets multiple times a day—short snippets of information in 140 characters or less and encourage retweets and interaction.
3. Google+	http://www.google.com/intl/en-US/+/business/	**What Google+ does:** Reaches out to local customers and connects with local businesses.
		How it works: Create a page, post statuses, post photos, engage with customers, and receive reviews and suggestions for improvement.
4. Pinterest	https://business.pinterest.com/en	**What Pinterest does:** Reaches mainly a female audience, but attracts customers through vivid, appealing pictures.
		How it works: Create a page, upload pictures, write captivating captions, and start pinning.
5. LinkedIn	http://smallbusiness.linkedin.com/	**What LinkedIn does:** Connects and markets B2B (Business-to-Business) companies, job recruiters, and job seekers.
		How it works: Create a business page, seek connections, follow other businesses, connect with individuals or companies.
6. Insta-gram	http://business.instagram.com/	**What Instagram does:** Promotes businesses that are related to lifestyle, food, fashion, and luxury brands.
		How it works: Post pictures that invite customers to follow you and attract them toward your products. Create an account and acquire followers.
7. Tumblr	https://www.tumblr.com/business	**What Tumblr does:** Reaches out to bloggers across the country and world; Promote trends and products for your business.
		How it works: Create an account, then share posts, pictures, reblogs, and other multimedia.
8. YouTube	https://www.youtube.com/yt/advertise/	**What YouTube does:** Reaches a vast audience that enjoys multimedia-processed information.
		How it works: Create a channel, use Google AdWords, and start posting videos.

9. Yelp	https://biz.yelp.com/	**What Yelp does:** Reaches a local audience for free.
		How it works: Create a page, post information, acquire customers, and receive reviews.
10. Four-square	http://business.foursquare.com/	**What Foursquare does:** Reaches a local audience and tracks people's location.
		How it works: Register your company/business then encourage people to "check in."

Below are more details on creating and managing a business page with the **top five social media platforms**. However, don't limit yourself to these if you're willing to try lesser-known social media platforms. As stated before, you never know what will work best for your business until you give it a shot.

Facebook: The Best Social Network For A Specific Audience

Brief History: Launched in 2004, Mark Zuckerberg and his college roommates at Harvard University created the early beginnings of Facebook in their dorm room. Facebook started out as a website that was only available to Harvard students. By 2006 it expanded to everyone who had an email address and was at least 13 years old.

Quick Facts: Facebook is the most popular social media platform on the Internet; more than a billion people use it daily. Apart from connecting with friends and sharing pictures, Facebook is one of the most effective marketing platforms for businesses. Although it doesn't give businesses the most traffic to their website, Facebook marketing receives the most customer sales and sign-ups. About 70 percent of marketers who used Facebook gained new customers, and around 47 percent of Americans said that Facebook influences their purchases more than any other social media platform. Studies have shown that the best time to update your status, post photos and interact with customers is between 1 p.m. to 4 p.m. daily.

Best Qualities and Benefits: You can maintain both a personal Facebook page and a business page. You can invite your personal friends to "like" your business page. Setting up your business page is free. To "boost" your page (which means promoting content to gain more likes and views of your business page), set your own budget (for as little as $5) and create advertisements to reach a specific audience. Most online advertising only reaches 38 percent of its intended audience; Facebook reaches 89 percent. Facebook advertisements are also best for raising awareness and generating sales.

Advertising on Facebook is very discreet. The more "likes" and interests a Facebook user clicks on Facebook, the more money Facebook makes and helps you advertise. For example, if your business sells makeup or other beauty products, then Facebook will find users who have "liked" makeup companies such as Sephora, MAC or Revlon and directly advertise your business to them. The more "likes" you as a user click on Facebook, the more opportunities it gives to businesses to advertise their products to you.

Terms To Become Familiar With:

- **Friend(s):** a person, or people, you connect with on Facebook; generally, only friends have direct access to personal information, photos, and status updates on a person's profile. Only personal page have "friends".

- **Like:** When you enjoy someone's content, you can "like" their text, status, video etc. to give them feedback. The more "likes" your content receives, the more potential customers you'll have. Note: when you "like"a page on Facebook, any status updates from the page will show up in your news feed.

- **News Feed:** This is the constantly updating list of stories in the middle of a user's home page. The news feed includes status updates, photos, videos, links, app activity and likes from people, pages and groups that you follow on Facebook. The news feed is what every user sees when he or she visits their own Facebook page. If you do not regularly post status updates, you will drop off your follower's news feeds.

- **Share:** If there is a photo, status, or video that shows up on your news feed, you have the ability to share it with other people. Encouraging your friends or followers to share your business page's content is key to promoting your business.

- **Pages:** what a business creates to list its information and share content with potential and current customers.

- **Timeline:** where all your shared content and information is located. Customers can visit your timeline to see what you've posted in the past.

Step-By-Step Guide:
You have many options to market your business on Facebook. Start by creating a business page for free. To create a page, there are a few easy steps to follow. You'll upload both a profile picture and a cover photo. Be aware of the guide-

lines when creating a page to make sure you look professional. Your profile picture is square and measures 160x160 pixels. Your cover photo is 851 pixels wide and 315 pixels tall. Next, you will be prompted to insert all your business' information. For a complete step-by-step guide, go to: www.facebook.com/business/overview.

Once you've created your page, invite friends to "like" it on Facebook or upload an approved email list to invite people to "like" the page and spread awareness about your business. After that, keep your page updated. Write status updates about events, new products, and offer sales—anything that will grab your audience's attention. Be sure to review Facebook's guidelines on contests and giveaways frequently as they are subject to change. Facebook also has a very comprehensive Help section where you can find the answer to most every question: www.facebook.com/help.

When **creating a Facebook ad**, you can specify what you want out of ads. You can create ads specifically to drive people to your own website. You can create ads to increase the number of people who like your page. You can pay a flat fee ($40) to boost a specific status update for 24 hours. You can post an "offer" (similar to a coupon) that users click to claim.

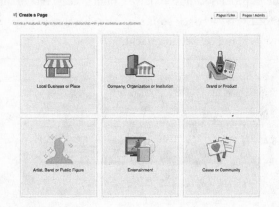

How Advertising Works: When you create any advertising on Facebook, you have the ability to target the ad very specifically. Facebook incorporates these top specific-targeting options: location, behavior, demographics, connections, interests, and custom/lookalike audiences. Below are their descriptions:

- **Location:** If you just opened a local sushi restaurant, then a Facebook ad can reach out to users who live in the area, have an interest in sushi, and may be interested in checking out your restaurant.

- **Behavior:** Facebook can take users' "behavior" and activities on Facebook into account and advertise relevant ads toward them.

- **Demographics:** If you own a pedicure spa and want to advertise to women from ages 18-60, then Facebook will promote your business to that targeted audience.

- **Connections:** You can advertise to people who are connected (friends of your friends) on Facebook.

- **Interests:** If Facebook users like or follow interests, such as TV shows, food, sports teams, and beauty products, then Facebook can advertise to potential customers who may have an interest in your business' products.

- **Custom/lookalike audiences:** If your business offers a product that might interest a Facebook user, then Facebook will advertise that product and your company to them.

Measure Analytics/Readjust: Measuring and tracking your marketing success is vital and will take time to properly adjust to have a high success rate of business improvement. There are **four ways** to track your results on Facebook: page results, ads manager, audience insights, and conversion tracking.

1) **Page Insights:** According to Facebook, "page insights are the analytics behind your Page that will give you information about who is connected to you and an overview of how they're responding to the content you share." After reviewing this information and seeing which posts received the most activity or least attention, you can plan better, more effective posts in the future.

2) **Ads Manager:** After you set up an ad campaign, you can use the ads manager to see if you are hitting your specific goals. You can also manage your ad accounts, billing information, and reports here.

3) **Audience Insights:** This tracking tool allows you to understand your audience better, see where they live, and what they are interested in.

4) **Conversion Tracking:** This is where you set up and manage conversion tracking pixels on your website. If you want your audience to do a specific action, such as attend an event or sign up for something, then you can track it here.

Facebook makes advertising fairly easy to use and cheap for your limited budget. Start out with a business page, and keep it updated. After a couple of months of running your business page, create an ad if you want to increase your presence. If you don't see immediate results, don't worry. Having a Facebook page will put your business on the map and give you great exposure to new customers.

Google+: The Best Social Network for Visibility and Reviews

History: Founded in 1998 by two Ph.D. students at Stanford University, Larry Page and Sergey Brin, Google was created "to organize the world's information and make it universally accessible and useful." Apart from being the world's most used search engine, Google created a social-networking service called Google+ in 2011 to help out local businesses. Google+ serves as a free business-marketing tool, as well as a social networking site for individuals.

Quick Facts: Once again, Google has taken the Internet by storm. Less than 3-years-old, Google+ is predicted to surpass Facebook within the next couple of years—or even months at the speed it's going. With already over one billion active Google+ accounts, the platform grows at a steady rate of 33 percent every year. Marketing with this platform would be a very smart move if you want to boost your local business' awareness. The best time to reach out to customers is between 9 a.m. to 11 a.m. every day.

Best Qualities and Benefits: Google+ makes it easy to access everything you need to promote your business. Your Google+ page helps build a loyal fan base and raise awareness. Customers can rate and review content to endorse your business by clicking on the "+1" button and re-sharing posts to their friends across the web. The more

+1's you receive, the more your page and content will pop up on searches related to your business, such as Google Maps and other Google searches.

Activities include starting conversations, finding people to connect with, responding to customer feedback, connecting face-to-face through video chats, following organizations and interests, and receiving quick updates on any type of information.

Google AdWords is another avenue of marketing your business on the web. The process is very simple: potential customers search for services or products on Google, your ad pops up on the search page, and you'll receive potential customers once they see your advertisement on the page.

The best way to attract customers is through using specific keywords that will direct people to your business, trying out multiple ads to see which one works best, and targeting specific locations. If you want strictly females who live in Jupiter, Florida, then you can advertise to just those women in that area. Google Adwords isn't free, but you can make a large profit of advertising from it for a very cheap price; it can be as little as $10 a day.

Let's take a look at an example: John's Crab Shack in Jupiter, Florida offers a full lobster dinner for 2 for $20 as opposed to $40 at other restaurant. The business wants to advertise to couples that live in Springfield, a youthful suburb nearby. In order to advertise to those residents, John's Crab Shack should go onto Google AdWords and use keywords such as "lobster dinner", "Springfield", "Jupiter", "crab" and "offers". When people search for these services on Google, John's Crab Shack's ad would pop up on the page and reach it's intended customers.

<u>**Terms to Become Familiar With:**</u>

- **Hangouts:** Video chat services that you can host to connect with customers, conduct business meetings, or just use to talk with an employee. You can group video chat up to 10 people and share photos as well.

- **+1:** The +1 button allows you to appreciate and rate which updates, videos and photos grab your attention. Every +1 you earn on shared content, the more that content will spread on the web.

- **Circles/Communities:** These serve as individualized communities of friends, connections, and family. You can choose to send updates, photos and videos to specific circles or to the public.

Step-By-Step Guide: Setting up your Google+ page is quite simple. You'll be asked to **create a Google account** (if you don't already have one) and then answer a few questions about your business.

After your page is set up, continuously update your content with event announcements, statuses, shared photos and videos, and whatever else you think your audience and customers may enjoy.

How Advertising Works: Basically, all the +1 content you share with customers, friends, and family will boost your unpaid search results on the web and improve your local visibility search for small businesses. Google doesn't take demographics or any personal information; you and your content advertise what your business has to offer: good products, service and a passion for what you do.

Measure Analytics/Readjust: Google+ makes it very easy for you to manage your page in one place. The Google+ dashboard allows you to update your information, track how much of your page is complete, monitor notifications/updates, share content with customers, video chat via Hangout, view top searches for your business on Google, see where your customers live, view social insights to track your posts' successes, and manage ads on your page. This all-in-one section makes it very easy for Internet marketing beginners to advertise their business successfully.

After viewing the most and least popular content on your page, adjust status updates, photos, videos and events to cater to your customers' wants and needs. If you're diligent and pay attention to what's working for your page, then your business will receive more customers and grow.

Twitter: The Best Social Network for Timely, Relevant News

<u>History:</u> Created in 2006 during a "daylong brainstorming session," Jack Dorsey, Evan Williams, Biz Stone, and Noah Glass started a news-oriented revolution. In just 140 characters, tweets can send out relevant information fast to reach anyone across the world. What started out as a news source soon turned into an effective marketing outlet for businesses whose customers crave rapid and relevant information.

<u>Quick Facts:</u> About 34 percent of marketers who used Twitter saw great results for their businesses. There are about 500 million tweets posted and 271 million users active on Twitter every day. Twitter was also the fastest growing social media platform from 2012 to 2013, with a follower growth of 44 percent. Getting information out fast on the web is Twitter's specialty. If your company provides content-based, relevant information, then Twitter is the best social media platform for you. The best time to tweet for the most effective results is between 1 p.m. to 3 p.m. every day.

<u>Best Qualities/Benefits:</u> Twitter's marketing motto, *"Connect your business to what people are talking about right now"*, keeps your business updated and relevant for customers. Users want to hear from businesses; According to Compete (a social media analytic company), Twitter users, on average, follow six or more brands. In today's society, customers want to engage with businesses on a personal level and stay up-to-date about their favorite products and services.

Terms to Become Familiar With:

- **Tweet, Tweeting:** A message or act of posting a message on Twitter containing 140 characters or less.

- **Retweet:** The button below a tweet that allows you to re-post that tweet on your timeline and news feed.

- **Mention:** Including a person's (i.e. @username) in a tweet so you can grab their attention. This will appear in your feed and is visible to all of your followers.

- **DM** (Direct Message): a private message that only you and the recipient can see

- **Reply:** responding back to a tweeter who mentioned you in a tweet

- **Hashtag:** The (#) symbol used before a word or phrase. This organizes conversations around a specific theme. (i.e. "Loving this new #book by #AtlanticPublishing" or "Working on my #marketing skills with #AtlanticPublishing.")

- **Trends:** A series of hashtags that many tweeters are tweeting at the moment. These tend to change on a hourly basis as a new trend occurs.

- **Follow, Follower, and Unfollow:** Subscribing to someone's tweets; a Twitter user follows your tweets; the act of unsubscribing to a user's tweets.

- **Timeline:** Where all your tweets show up, archived.

Step-By-Step Guide: There are a couple of ways to market on Twitter: You can use Twitter ads or have a tweet activity board.

First, make a Twitter account for your business then engage followers with great tweets, trends, and relevant information. Choose your Twitter name carefully. You want it to be relevant and easy to remember. Twitter also has a profile picture and a header photo image. Your header photo is the main image across the top of the screen and should be 1500 pixels wide x 500 pixels high, and no larger than 5MB. Your Profile Photo, or Avatar is square 400×400 minimum with a 2MB maximum. Twitter also has a comprehensive help section that will walk you through setting up an account. Visit **https://support.twitter.com/articles/127871-customizing-your-profile** or **https://business.twitter.com**.

You can also use Twitter promotional tactics to launch a product, drive sales or increase brand awareness, such as getting followers to retweet a specific tweet in order to receive a gift or tweet exclusive offers and deals that customers can only learn about through Twitter.

When it comes to advertising on Twitter, there are different options. If you use an objective-based campaign, then you can focus on receiving more followers, drawing people to your website or store, tweeters engaging in your posts, or how many new people install your business' app. You can specify what you want out of marketing and can always switch it up whenever you want.

How Advertising Works: If you pay for a Twitter ad, then you can promote and recommend your account to a surplus of users for a certain amount of time. Twitter will incorporate users' information, interests, and followed accounts and refer them to follow you if they may have a potential interest in your business.

Measure Analytics/Readjust: In your Twitter dashboard, you can look at the number of follows, mentions, activity that followers like, tweets displayed, and what people are engaging in on a monthly basis. Whether it's retweets or a promoted account, marketing on Twitter will benefit your business in one (targeted) way or another.

Pinterest: The Best Social Network for Image-Related Content

History: Started in 2010 by founder Ben Silbermann, Pinterest became a social image bookmarking system that surprisingly revolutionized business marketing on the web. Pinterest has quickly become one of the best ways to attract customers to your website and purchase your products.

Quick Facts: According to recent surveys, Pinterest users are 79 percent more likely to purchase items they saw 'pinned' on Pinterest boards versus products Facebook users saw on their news feeds. Consumers are more responsive to image-based ads, and Pinterest makes it so easy to connect a customer to a buyer's page. The best times to post pictures and share content on Pinterest are between 2 p.m. to 4 p.m. and 8 p.m. to 1 a.m. daily. If your business sells items or image-related products, then be sure to check out Pinterest.

Best Qualities and Benefits: Drawing people to your website has better rates on Pinterest than Twitter or Facebook. Shoppers referred to a website through Pinterest spend 10 percent more than those referred by other sources. A marketing tactic called Promoted Pins will launch soon and help you reach more people on Pinterest, expanding your audience. You can add the "Pin it" button to the content on your website so customers can share your photo with other people. Most consumers are drawn toward image-based content, so marketing on Pinterest will definitely benefit your business.

Terms to Become Familiar With:

- **Pin, Pinning:** an image that you can share to followers; the act of putting an image on a board. You can pin an image from any web site or upload your own photos or images. You can also write a description for each pin.

- **Board:** here your pins are saved and available for other users to view

- **Follow:** On Pinterest, you can follow every board a person or business creates. Or you can follow just an individual board.

- **Home page:** This shows all the current "pins" every person you follow "pins". The more people you follow the more your home page changes as new pins are added.

- **Your Profile and Pins:** This is a page that displays your profile picture, a short description, your website and a summary of all your boards.

Step-By-Step Guide: Create a business page on Pinterest, follow current and potential customers then add high-quality images with rich, short descriptions that followers will love. You can sign into Pinterest using a Twitter or Facebook account and when you "pin" something have it posted automatically to Twitter or Facebook. Pinterest provides backlinks to your website, boosts your business on Google searches, and increases visibility of your business.

Besides pinning your own content, share other pinners' content as well (that's relevant to your business). Customers will enjoy the various content you share besides your own.

Before becoming a pin guru, think about what your audience likes then create boards that flow cohesively together. It's important to have content that works seamlessly for you and your customers.

How Advertising Works: For a free to cheap price, you promote your products on Pinterest and attract customers to click on the image and head toward your website. It's a simplified process that gets straight to the point: expanding your audience, making a profit, and raising awareness of your business.

Measure Analytics/Readjust: In regards to Pinterest analytics, you can track how many people are pinning from your website, view your pins, and click on your content. Learning from what your customers like will benefit what you and your business should produce. It's important to study pinners' activity because it's the best way to become successful.

A note about Pinterest activity: While current Pinterest activity shows up in your followers' home page or news feed, Pinterest also has a search function where users search for specific subjects. Searches are returned by either individual pin, board, or pinner. People will repin your pins, generating awareness from searches. As long as you spend some time setting up your Pinterest, creating boards and initially making pins, Pinterest is the one social media site you do NOT need to maintain daily for it to generate activity.

LinkedIn: The Best Social Network for Business-To-Business Growth

History: Launched in 2003, a combined group of SocialNet and PayPal workers thought of a business-networking idea and made it come true. Within a decade, this social networking website reached 225 million members. LinkedIn became an avenue for businesses to grow together and have easy access to resumes. Although it doesn't compete with social media platforms, LinkedIn connects businesses better than any other social networking site.

Quick Facts: LinkedIn has become one of the best business-to-business (B2B) networking websites on the Internet and is considered the world's largest audience of influential and affluent professionals all in one place. Marketing your company on LinkedIn brings in opportunities for new business connections as well as opening you up to professionals who can turn into huge assets for your business. LinkedIn is a great resource for finding talented individuals who are looking for new opportunities. It connects employers and employees with the added bonus of having readily accessible work histories, resumes and skill reviews. The best times to post updates and make new connections are between 7 a.m. and 9 a.m., and 5 p.m. and 6 p.m.

Best Qualities/Benefits: If your business focuses on B2B connections or is having a push for new talent, LinkedIn is your best site to use. You can find and connect with businesses easily.

Term to Become Familiar With:

- **Connections:** potential customers, friends, or co-workers who have access to your personal page and information, as well as their own page

- **InMail:** Used to make a connection with someone who you do not know or just to start a conversation, the in-system mailing network allows you to manage all communications from your home page

- **Recommendation:** LinkedIn allows you to recommend businesses or individuals by writing a review, much in the way of a Yelp! review.

- **Creative Portfolio Displays:** For those professions that deal in the creation or original content, the Creative Portfolio Display provides a professional space to organize and present finished products

Step-By-Step Guide: Make a company page. Your page raises your business' awareness, promotes career opportunities, and reaches out to potential customers and employees. When you interact with people on your page, you educate them about what you do, building trust and brand awareness. You can also target people by job function and industry as well as university, location, and current employer. By doing this, you can build targeted advertisements and gather specific information about

your desired audience. To this end, the polls feature provides tools for gathering actionable data.

How Advertising Works: LinkedIn advertises your business based off of demographics, interests, job descriptions and other categories. If you're looking for a very specific audience, then your LinkedIn company page will get the job done. People also join groups on LinkedIn based on these categories and you can target your ads just towards a specific interest group.

Measure Analytics/Readjust: LinkedIn helps you see if you are reaching out to the right people and doing it the most effective way. Take some time to review what works and doesn't work for your page so you can expand your potential customer audience.

You may feel overwhelmed by the amount of information you just covered, but don't let it get to you. These social media platforms weren't built into empires in one day, so try one networking site at a time. Spend some time familiarizing your self with a site, review the help sections, and talk to other business owners who maintain social media sites. It takes time to build up a customer following because it takes time to build trusting relationships. Social media is meant to benefit you and your business. You can also hire a social media consultant to get you started or do some one-on-one training. Check out **www.elance.com** or **www.fiverr.com/categories/online-marketing**.

Email Marketing: Why It's Still Effective

Before social media exploded onto the scene, email marketing was one of the few ways to advertise via the web. Even though social media has caught many businesses' attentions, a 2013 study from a predictive analytics firm stated that customer sales have quadrupled within the last five years just from **email marketing**.

Building an email list is the most important factor of effective email marketing. Having current and potential customers' consent is also important; they must be *fully aware* that they gave you their email and/or agreed to receive emails from your business. You can obtain email addresses through customers who sign up for newsletters and want to know about exclusive offers or hear from you about upcoming products and services.

While social media is great for connecting with customers, it doesn't bring in as many sales as email marketing does. People check their emails multiple times throughout the day, which makes email marketing a perfect avenue for you to get your information into personal emails. Email marketing is not only a great way to get in touch with your targeted audience, but it's also a cheap way to market your company. Below are fast and easy steps to start marketing via email:

- **Define your targeted audience**
- **Determine your content**
- **Determine email frequency and goals**
- **Observe results**
- **Repeat or start over**

It may take some time to figure out how to effectively use email marketing, but everything takes time and practice. Using email as a marketing source will benefit your business by expanding its presence, reaching potential clients, and stretching your advertising abilities.

The biggest mistake that business owners make when advertising through email marketing is not capturing customers' information. The purpose of advertising includes understanding your customers better (i.e. through likes, dislikes, and needs) and reaching out to them. Study your email marketing results to see how to market the most effectively to your intended audience.

Web Marketing: Why It's A Tried-And-True Method

With all the social media hype and email-marketing strategies, business owners forget about traditional web-marketing strategies, such as search engine optimization (SEO) and banner advertising. Although some of these marketing methods aren't as new and appealing as social media, you should still check them out and see if any marketing tool will help you out.

Search engines, such as Google, Bing, and Yahoo, control your website's presence. The more your business is known, the more customers you'll attract. Below are short histories of the founding search-engine powerhouses:

- **Google:** Started in 1996 as a research project by Larry Page and Sergey Brin, the most powerful company and search engine has changed education, socialization, business growth and life itself. As the leader in SEO, Google controls and changes algorithms that affect Internet searches. It's important to keep up-to-date with how Google conducts its business strategies.

- **Bing** (previously known as Live Search, Windows Live Search, and MSN Search) is a web search engine launched by Microsoft in 2009. The engine has used "Tiger", a new index-serving technology, since 2011. Bing is the second largest search engine after Google.

- **Yahoo!** Is another large web search engine founded by David Filo and Jerry Yang in the early 1990's. After going through a numerous CEOs and losing tons of money, Yahoo! still remains one of the most used search engines.

SEO is the process that affects the visibility of a website or web page in a search engine's 'natural' (or unpaid) search results. In other words, it's a search for businesses or companies where their website pops up pretty high on the search page without paying for the website to bump up higher.

Search engines create mathematical algorithms that affect your presence on the web. Typically, search engines only give weight to the **first 200 words** on your website, article or other content. For instance, if you have many links and most used words on your website, there's a higher likelihood for Internet surfers (or customers) to stumble upon your website. Below is a list of great SEO content characteristics to include on your website and in your content:

- **Words matter:** The more popular words you hashtag or use on your website, the more likely your website will pop up higher on the web. There are constantly new articles that keep marketers and advertisers aware of trending SEO words and hashtags. Keeping up with SEO trends can be difficult at first, but after awhile you'll know which content to add and tag. However, only include keywords and hashtags that pertain to your subject or content; otherwise you may be penalized for 'keyword stuffing' (incorporating too many keywords or random words that have nothing to do with your content).

- **Titles matter:** If your website or articles include popular keywords, then the search engine will give weight to your content. Creating great titles is very important and should follow the current trends on the Internet and in the business world.

- **Links matter:** Including good sources and active links on your website or article gives more weight to your presence, too. However, if you overload on faulty or purchased links, then search engines will penalize it against you. Always balance the use of links and make sure they lead to solid content.

- **Reputation:** Even though reputation is important, this doesn't mean that small, up-and-coming businesses don't have a chance to widen their web presence. Yes, Amazon, the New York Times and ESPN are going to pop up before your website, but search engines have created a special method to boost everyone's web presence, if you learn how to use SEO. Your business may start off at the low end of the totem pole, but you'll expand and work your way up to the top with perseverance.

Another great web marketing tool is **Groupon**. Groupon is "a deal-of-the-day website that features discounted gift certificates that are used at local or national businesses and companies". About 650,000 businesses have marketed on Groupon, and there have been 600 million deals purchased so far. Seventy-eight percent of customers said they wouldn't have purchased an item or service if they hadn't first bought it off Groupon.

The downside of Groupon is that you must offer your service for half the price (50 percent off) and then pay Groupon half of the amount you received. So you don't make much of an income, but it at least advertises your business and attracts people to try your products and services. A good way to make Groupon customers repeat customer is by providing the best possible service on the day of. To do this, add on more staff on the day of the promotion. Try to make it a day you are not normally busy. Give your staff a "Groupon Bonus" for every customer they help. At they end of the day, collect email addresses so you can continue to market to them. This strategy ensures that you provide the best service possible and that your staff enjoys the busier days.

Some people are "bargain hunters" and don't come back a second time, but sometimes you can attract many customers and have successful results. **LivingSocial** is another great website like Groupon that offers advertising services on the web for your business.

After learning about the different types of Internet marketing, it's up to you to decide where to go from here. If you're still unsure about which marketing avenue to pursue, hire an experienced marketer, public relations specialist, or advertiser to help you get started. It truly is a great investment for your company. Advertisers know how to brand your company exceedingly well and will expand your presence. Getting experienced advice will help you out so much. Make smart decisions before jumping into the online marketing melee.

CASE STUDY:
PAUL'S PASTRY TRUCK

Who: Paul's Pastry Truck
What: Food Truck (Pastry Specialty)
Where: Nowhere, FL (and surrounding area)
Details: New local business with a twist
        (and tight budget)

Meet Paul—the new owner of a pastry business in Nowhere, Florida. At age 30, he's finally made his dream come true: he is opening up the first pastry food truck in town. He plans on officially opening in a month and is willing to commute up to 20 miles near Nowhere, Florida for food-truck shows and events. Paul wants to advertise on social media for his business but is on a tight budget. So where should he start?

First, let's take a look at Paul's situation to see which social media platforms would work best for him:

- Paul has a business that offers unique services: He can travel to customers with his food truck and provide delicious pastries anytime, anywhere.

- He sells pastry and some other pet geared items: his targeted audience ranges from kids to adults; every age group enjoys pastries.

- Paul's Pastry Truck is a new local business: locals may want to try out a new place and support local businesses in their community.

After looking over his services and situation, now he can determine which social media platforms would be best to use. Here are four that he can choose from:

1. **Twitter:** For location updates, Twitter is the perfect social media outlet for Paul's Pastry Truck. With Twitter, Paul can send fast updates about where his next events will be, expand his audience by encouraging retweets for free cookies, and include exclusive information that can only be found on Twitter. Having a Twitter account could greatly boost the pastry truck's audience and bring in business.

2. **Google+:** Having a Google+ page would broaden the pastry truck's local audience very quickly. By encouraging customers to rate and review different pastries, sharing exclusive information with customers, and give location updates, Google+ can boost Paul's audience effectively.

3. **Pinterest:** This could be another successful social media outlet for Paul. By displaying high-quality pictures on Pinterest and adding short, eye-catching descriptions, Paul could attract people to his website and draw more customers to purchase his cookies and cupcakes.

4. **Facebook:** Having a Facebook page would boost Paul's business by posting updates, upcoming events, captivating pictures, and any other enticing media. He can invite his friends to like his page and they can share it with their friends in return.

In a perfect world, Paul would be able to handle all four social media outlets, but he just simply doesn't have enough time to manage them all. He should use two at the most until he has more time and learns how to balance cooking, interacting with customers, and managing his social media accounts. The best two social media outlets Paul should use are Twitter and Google+. These accounts will benefit him the best through timely updates, reaching a local audience, and gaining new customers quickly.

Paul thought about using email and web marketing, but he doesn't think it would work as well as using social media platforms. His intended audience would ignore newsletters about new pastries in their inbox, and selling his product on Groupon wouldn't benefit his income; so Twitter and Google+ are Paul's best options for marketing.

After a month or two, Paul should greatly benefit from both social media platforms. His business will start off very well and make him a respected competitor with other pastry businesses in the area, even though he just launched his business.

Conclusion

Internet marketing is a growing trend that's not dying. For some business owners, getting started is the hardest part. You may still be unsure how it all works, but taking small steps will bring in new customers, amp up your business' presence, and make you more aware of how the business world works now.

Search the web for more tips to start a social media account, and utilize email marketing tricks or whatever else you want to try.

The Internet is meant to be a tool for boosting your business. Take advantage of this free commodity and make your business known.

Quick Summary of Strategies

1) Create your own **website**. It's quick, easy, and free. (i.e Wix, Weebly, Wordpress, Tumblr, Blogger, and Sitey)

2) **Read, Read, Read:** Discover what social media platforms, email marketing tactics and web marketing tools you want to use.

3) Set up your **social media** accounts. Don't keep putting it off; start today.

4) Get down to business: **update** statuses, **share** photos, make **events**, host **contests**, and **interact** with your potential customers.

5) Study your **results**: After a month or two, look over what followers like and dislike. Learn from mistakes and celebrate what's worked for your business.

6) **Repeat:** Internet marketing is continuously changing. Stay in the know about satisfying your customers and expanding your business.

What is Traditional Marketing?

Hiring Marketing Experts

Outsourcing your marketing to an advertising or PR agency can be an excellent business investment. For an hourly rate, either a fee or percentage of your advertising purchases, a full-service agency can help you:

- Create your image – design a logo, mascot or visual message.

- Construct an advertising message – create a slogan that represents your business and reminds potential customers of what you do.

- Develop advertising campaigns – recommend the right media to reach your ideal customer.

- Design ads for print, TV, radio, Facebook, Twitter – produce print and electronic ads.

- Create positive buzz – develop public relations activities.

- Make media buys – negotiate and purchase ad time or space.

As a small businessperson, your budget may restrict you from hiring a full-service advertising agency. Hiring independent marketing consultants and freelance business communication writers can be cost-effective alternatives.

When hiring marketing experts, look for:

- Experience with food service businesses.

- Smaller firms where you will be working with experienced marketers instead of underlings.

- Well-developed proposals.

- Willingness to work within your set budget.

Remember: Investing in the services of an advertising agency or marketing consultant can pay for itself. They can steer you away from costly mistakes, create a consistent message, and negotiate lower ad rates.

Do-It-Yourself Marketing

Millions of entrepreneurs successfully advertise and promote their business. They make wise buying decisions, track their ads' effectiveness and grow their business. Conversely, others overspend on promotions that don't increase sales; fail to create a single image or message; and worst of all – don't increase their sales.

Should you handle your own marketing?

- Are you interested in learning about marketing?

- Is it a good use of your time?

- Can you be objective?

- Can handle aggressive ad salespeople?

- Do you have a creative eye?

As you can see from our examples above, marketing your business includes the details of daily business life. These are all your responsibility; however, the more complex and costly marketing efforts can be successfully handled in-house also.

If you have no prior marketing experience, you can learn and implement practical ways to let people know about your business. Below are some resources to help you set your plan in motion.

- The U.S. Postal Service offers direct mail services and advice for small businesses. Visit **www.usps.com/business/business-shipping.htm** to learn more.

- Do-it-yourself email marketing can be a great way to reach people. Constant Contact (www.constantcontact.com) provides user-friendly services. Email Factory (www.emailfactory.me) and Got Marketing (**www. gotmarketing.com**) are other cost-effective e-mail services.

- Learn about specific marketing techniques at Market It write (**www. marketitwrite.com**).

- Explore creative marketing ideas at Idea Site for Businesses (**www. ideasiteforbusiness.com**).

- Read *The Handbook to Retail Bakery Marketing: An Introduction to Merchandising* published by: CAIMA, 32 Dromore Crescent, Hamilton, Ontario, Canada, L8S 4A9, 905-528-8371.

Desktop Publishing Applications and Ideas

There are hundreds of reasons to own and utilize a computer in your work as a food service manager. The computer, if utilized effectively, will save you an enormous amount of time and money. Here are just a few ideas for desktop publishing:

Print your own customer and/or employee newsletters, table tents, discount coupons, direct mail flyers or postcards, menus, business cards, employee-of-the-month certificates, customer gift certificates, advertising posters, employee manuals, office stationery and newsletters.

If you are going to print your documents on your own laser or inkjet printer, Microsoft Publisher® is a simple to learn program that includes templates and some basic artwork. Microsoft's website also has free enhancements for Publisher.

If you want more control and perhaps need to have a professional printer handle the finished design, Adobe PageMaker® is a full-featured desktop publisher and the newest versions include an assortment of pre-designed templates for common items. You can use these templates in the provided color schemes and artwork or modify them to suit your needs.

Beyond the basics of advertising and menu creation, we recommend hiring a graphic designer or printer with an in-house design staff. Don't spend unnecessary hours when they can do it pronto! Doing it yourself has a point of diminishing returns – your time should be spent where it can make you the most money.

Do I Need A Website?

Each day, billions of people around the globe use the web for work, play, shopping, and research, and you don't want to miss out on that enormous source of potential customers. Your business will be visible to your neighbor, a French homemaker and a Japanese businessman.

Perhaps your goal is to serve the freshest sourdough bread in your city or to create elaborate wedding cakes; so why would you care about visibility beyond your community? Because the world shrinks every day and the possibilities are unlimited – in whom you can reach, what you can sell or who will find their way to your door. No other marketing tool can provide such comprehensive coverage for such a nominal investment.

Your website is your full-color brochure where you can tell people:

- Who you are – ***Bread bakers extraordinaire***

- What you make – ***We specialize in rustic breads***.

- Who you serve – ***Serving Old Town Tustin residents since 1988***.

- Where you are located – ***At the corner of Red Hill and Spruce***.

- When you are open – ***Serving hot bread from 7:30 a.m. through 7:00 p.m. Monday through Saturday***.

- Why they should buy from you – ***We use only the freshest organic ingredients***.

- How to place your special order – ***Use our online order system and your special order will be waiting for you!***

The Internet offers marketing benefits beyond those of a simple business card. Your site can:

1. Be updated quickly without any "waste." You don't have brochures to toss when you introduce a new signature cake. Make their mouth water with beautiful pictures!

2. Be your electronic menu – Order your Challah today.

3. Grow it as you grow – it can be whatever size you want.

4. Be interactive – where people are given a chance to respond directly to your information.

5. Have lasting value – you can share cake-serving techniques.

6. Build community spirit – you can announce and run charity events through your site, promote community activities, and provide a community forum.

7. Sell to your specialties to "foodies" around the globe with e-commerce features.

Your website can also be a great way to communicate with your employees. A password-protected section can feature a company newsletter, explain benefits or post work schedules.

Reaching customers within your own community can also be done on the web via digital cities, online city guides and other food or community-oriented sites where you can place an advertisement, free directory listing or link your website. Chambers of Commerce, virtual travel guides, online wedding planner services and guide sites such as About.com at **www.about.com** are all possibilities.

Use the checklist of potential advantages below and see for yourself. Place a checkmark next to each trait that would serve your business:

- Reach a global market.

- Gather marketing information.

- Analyze and evaluate marketing information.

- Generate additional sales.

- Establish more frequent and meaningful communication with customers and employees.

- Supplement employee training through electronic updates and bulletins.

- Broadcast press releases.

- Submit invoices and expenses more quickly.

- Identify and solicit prospective employees.

- Provide immediate access to your catalog.

- Permit customers to place orders electronically.

- Reduce costs of goods sold through reduced personnel.

What To Put On Your Website

What kind of information can a bakery put on the web? Remember that the site should reflect your bakery's personality. If you think of your bakery as classically elegant – then your site should have that appearance. If your bakery's décor is bright yellow and playful – then your site should have yellow accents and have a playful feel. Here are some ideas of what you can include on your business website.

- A picture is truly worth a thousand words. Don't just tell them you bake mouth-watering cakes. Show them –
  - o how inviting your dining area is.
  - o a sparkling display case filled with enticing pastries.
  - o your cheerful staff at work.
  - o how you make the delicate frosting flowers.
  - o huge steaming cups of coffee.
  - o ideas for special event cakes.

- News, events and specials. Develop a web-based newsletter to share your exciting news.

- Menus. These aren't just basic menus, but menus with full-color photographs of each entree!

- Directions. Enter your address, and you get door-to-door directions from your home to the bakery.

- Non-bakery products for sale. Cake plates, wedding cake tops, serving sets, T-shirts or posters can be purchased online.

- History. Every town has a history. If your bakery or your building has an interesting history – share your story.

- Area Attractions. Sell your bakery *and* your local community to the website visitor.

The opportunities are endless. Be imaginative!

How Do You Get An Effective Website?

You can build it yourself or you can hire an experienced web development company. However, remember that just because a program is convenient, doesn't mean that you should spend your time becoming a web designer.

You would need a unique domain name if you want to have your own web site. To determine whether your bakery can be used as a dot com (.com) or dot net (.net), visit the Whois database of web names at Network Solutions at **www.networksolu-tions.com/whois/index.jsp**. Due to the tremendous growth of the web, you may find your business name is already registered as a dot com. Try the dot net extension also. Variations such as adding your city name may be available. Avoid long names as they are hard to put on a business card and even harder for people to remember. Dashes should likewise be avoided as they can be confusing. Once you have found an available name, you can register it at Network Solutions (**www.networksolutions. com**) or one of the many registration companies on the web. Fees vary so shop around and beware of companies that don't allow you to register yourself as the "owner" and as the "administrator."

When you have registered your domain name, it needs to be "pointed" somewhere on the Web. Some registration companies offer "free" camping until you have a website set up with a hosting company. You can create unique e-mail addresses such as **betty baker@sweettreats.com** or **orderdept@weddingcakes.com**. Using different e-mail addresses on advertising can be a good way to tell how people found out about your bakery and the effectiveness of a specific ad campaign.

Remember that:

1. Your site must look professional – a "homemade" site can reflect poorly on your business.

2. Your site must "work" – the mechanics behind the site and the navigational system are what makes it a pleasant visit for your customers and potential customers. Broken links or hard to operation ordering systems are real turnoffs.

3. Your site needs to be search engine friendly – since 75 percent of all online activity comes from finding a site via a search engine – your site must be easy to find. Your design and keyword filled copy are important components.

4. Your site needs sales-oriented copy. If words fail you, you'll need a copywriter or at the very least someone with a flair for writing and excellent grammar and spelling skills.

5. Your time may be better spent on other bakery activities. Avoid the entrepreneur trap of taking on too many tasks. Do what you do best and hire others to support you.

Hiring Web Pros

- Use the web to find the right web developer for your business.
  - o Search on "web design [your city name]" to find locals who can handle the job or search "bakery [or restaurant] web design" for people across the country with experience in designing bakery or food-service sites.

- Look at other bakery sites.
  - o If you see one that you like, contact the webmaster (usually listed at the bottom of the page) and ask who created their site.
  - o Make note of the bakery sites you visit. Share what you like and what you don't with your Web designer.

- Review web developer portfolios and site samples.
  - o Are they attractive?
  - o Do they function properly?
  - o Are they filled with annoying sounds and whizzing pictures?
  - o Do you know immediately upon visiting the site what they do.

Some Words Of Caution

- Don't overlook the little details. Web users have become more savvy and have big expectations – even for a local bakery site.

- Invest your time and money wisely. There is always a point of diminishing returns and inflated expectations.

- Build a site that can grow with you.

- Keep in mind the "hidden costs." Most developers don't include Web site hosting, domain-name registration and renewal, support and continued development services after site completion.

- Make sure you promote your site. A site is worthless if no one knows it exists. Search engine registration is a critical part of a successful website.

A well-constructed website is an investment, not an expense. Integrate it with your newspaper, Yellow Page and other offline advertising efforts to maximize its returns.

Find Support with Suppliers

Suppliers can also be a huge ally, because the more business *you* do, the more orders you send *them* for their products or services. Manufacturers often have co-op advertising programs where dollars are available for promoting the use of their products by your business. These funds can be used to create local PR campaigns and advertising in exchange for mentioning the company brand names. To locate co-op programs and confirm your eligibility, speak with wholesalers, distributors and other vendors about all available programs. Be aware – these "gifts" come with rules and regulations. Be certain to read them carefully, document your eligibility and keep good records.

Some publications (telephone books, directories, small newspapers, etc.) will provide creative services when you purchase an ad. This can save money; however, you may have limited control over the design and your ad may end up looking just like everyone else's.

Band Together

Another practical way to share the cost of advertising is to create an association of complementary businesses to produce omnibus ads. You can gather together businesses in your neighborhood, mall or business center and create a network for other "food" businesses such as a gourmet cook shop, a produce market, a wine store and a fish market. The businesses can be close for "walk-in" campaigns or located around town with similar customer demographics.

- Share the cost of a half-page newspaper ad with other businesses in your strip mall or with other wedding service providers. Dividing the cost is frequently less expensive than purchasing your "share" separately.

- Create in-store promotions for each other. Think of creative ways to cross-market your ad partner's products and do the same for your desserts in their shops.

- Develop a punch card frequent buyer program with an ad partner. Think of ideas like receive a bouquet of flowers with every 12 loaves of bread purchased!

- Provide the refreshments for open houses and other special events. Create an attractive display and always have a sign that holds business cards.

- Hold a "sidewalk" sale with neighboring businesses. Provide food samples, plenty of seating and discount coupons to encourage people to return to your bakery later.

Your local chamber of commerce and other business associations are great resources for connecting with and building your marketing network.

Investing in Customers

So how much should you spend to acquire a new customer? There is no exact right or wrong answer. The key is to compare this cost against the **lifetime value** of a customer. If your new customer came in, bought a bagel and never came back, the $8 would be a very poor investment. That's why customer retention is so important (see Chapter 7 on delighting your customers).

Lifetime Value

Let's use a simple example to help you realize the potential value of Sarah Brown. Sarah and her husband have two growing boys. Sarah purchases two loaves of bread, one cake, two-dozen cookies and a dozen bagels each week. Sarah will live in your neighborhood for 5 years and visit your bakery every Friday afternoon after work.

	2 Bread Loaves	1 Cake	2 dozen Cookies	1 dozen Bagels	Weekly Total
Weekly Purchases	$5.90	$14.95	$6.50	$7.90	$35.25
Annual Purchases*	$295.00	$747.50	$325.00	$395.00	$1762.50
5-year "life"	$1475.00	$3737.50	$1625.00	$1975.00	**$8,812.50**

* Less 2-week vacation

The $8 customer acquisition would be repaid 1,100 times over! Of course, you spent money throughout the five years Sarah shopped at your bakery to *keep her as a customer* so your true return wouldn't be as dramatic; but from the example you can see that advertising can be a great investment.

Calculating Your Customer Acquisition Cost

If you are a new business, every customer is new and you'll know your exact advertising costs. At the end of your first 30 days, divide your customer count (assume individual transactions) by your actual advertising expenses[1] from pre-opening to the end of the 30-day cycle. In subsequent months, you'll compare the next 30-days with the prior. After a year, you'll be able to compare one 30-day period to the same period in the prior year. Some people use 13-month cycles for these calculations gives them results that are more accurate.

If you are taking over someone else's operation, you'll have the prior owner's financial data to use as a benchmark. If they did not keep detailed records, you can use some sales per customer factors to estimate the customer acquisition cost benchmark.

Below are some resources and tips on calculating your customer acquisition cost.

- These assumptions don't take into account the residual benefit you receive from the prior owner's good reputation along with prior and ongoing word of mouth advertising, PR or formal advertising.

- RJ Metrics offers a **lifetime value of a customer** calculator at **http://customerlifetimevalue.co.**

- Data gathered on a daily and/or monthly basis must be compared to like periods.

The purpose of this exercise goes beyond decision-making and budgeting – it should also help you *feel more comfortable* in investing in solid long-term advertising campaigns. In addition, recognizing the long-term value of a customer reinforces the wisdom of giving a customer a free cake to satisfy a problem. Your cost of $6 is just a fraction of what they give you!

Customer Costs

Since your goal is repeat business and, hopefully, people will start coming in regularly during your first month of operation, your "new customer" data (see above) won't be accurate. If all your business were custom, one-to-one sales, you'd easily tally who is new and who is a repeat customer. However, a typical retail outlet will have to make estimates and assumptions.

---

[1] You should allocate a percentage of your annual advertising buys over the 12-month accounting cycle. If you purchase a $1,000 ad in July but it doesn't appear in print until September, the expense would be used in September calculations. In other words, use the publication/air date as opposed to contract or payment dates.

The basic formula for computing your customer acquisition cost is your total advertising costs divided by the number of new customers. Let's say you spent $800 in June (not your first month) and the number of individual sales (your customers) rose by 100 from the prior June. Your June cost to acquire 100 new customers is $8. In the real world, you may have only had 25 new customers. However, your great products, excellent service and ongoing advertising meant that you rang up an extra 75 sales.

Advertising

Advertising is a direct activity where you purchase time or space to present your message. You purchase time for cable ads or radio spots and space for newspaper display ads or a Web site. Advertising is often a broad activity where you reach out to a huge audience in hope of capturing the attention (and money) of a few potential customers. There are two major factors you'll use to calculate your advertising ROI (return on investment). The first is the number of customers you acquire and the second is the acquisition cost of each new customer.

By selecting advertising that targets your ideal customer demographics, you'll increase your success rate. For example if you have a bakery specializing in wedding cakes, you wouldn't advertise in a PTA newsletter but you would have a booth at the local Wedding Fair. Sure, you might find a wedding cake customer who reads the newsletter, but you'd have much better success if you talk directly with future brides. The more people you connect with that match your "ideal customer" profiles, the more cost-effective your advertising.

Calculating your advertising's potential value can be done by:

- Dividing the number of impressions or viewings (impressions means copies printed and viewings are for visual mediums such as billboards and the Web) by the cost of the ad.

    o   An example – you purchase a small ad in the local community newspaper for $100. Their circulation is 10,000. Divide $100 by 10,000 for a $.01 cost per impression.

    o   Compare that with – you purchase a regional magazine ad for $800. Their circulation is 200,000. Divide $250 by 200,000 for a $.004 cost per impression.

    o   Assuming all other factors (demographics, distribution, etc.) were equal, the best value is the magazine ad.

o   However, if your $100 newspaper ad will be going into the food section, a special Wedding issue or other "targeted" issue, that ad would be a better value for your bakery.

Your Advertising Effectiveness

For a retail operation, it can be very difficult to figure out which advertising brought new customers in. Did they find you in the Yellow Pages® or see your newspaper ad? You need to build in ways to gather this information into your campaigns whenever possible. Without knowing how they found you, it becomes more difficult to eliminate the non-productive advertising and increase the productive types. Below are a few ways to see which ads are working for you:

1. Ask your customers how they found you.
   a. On the web, include a field on forms.
   b. Over the counter, periodically have staff ask people as they ring up their order. Keep a simple running tally.
   c. Include a postage paid reply card in to-go packaging.
   d. Include a reply card that when brought back *completed*, they get a discount or a freebie.
   e. Ask and record when taking phone or face-to-face custom orders.

2. Set up automatic data gathering mechanisms.
   a. Unique email addresses for specific promotions.
   b. Unique phone numbers for specific promotions.
   c. Coupons with codes indicating when and where they were printed and/or distributed.

Taking time to learn what ads are drawing people in the best will help you make critical budgeting decisions. You won't have to assume that your billboards are working when in fact it is the door hangers.

Some Ads Take Time

Some ads bring people into your bakery immediately; others plant a seed for future reference. You will be attracting two types of customers: 1) those with an ongoing need for your product (a weekly loaf of bread); and 2) those with a special need (a wedding or birthday cake). By adding a sense of urgency to your ads, you'll bring in the first group. The second group fits into the future reference category because unless your ad happens to appear just when they are ready to buy.

To add a sense of urgency to your advertising:

1. **Always include an end date for all offerings.** For ads that are on TV, radio or newspapers, the special should end in a week or less. For magazines with a longer "shelf life," a 30-day or longer offer would be more appropriate. Always be aware of print and pull dates. You'll turn off potential customers if your offer expired before the magazine was removed from the newsstand.

2. **Include a call to action.** Tell them to drop by today, to bring in the coupon, to check out your website, to try your sample and register to win!

Repetition, Repetition, Repetition

If you have to choose a smaller size ad to have it appear more often, opt for a greater frequency. Many advertising experts believe that it takes three to seven times of being exposed to your offer before people make a purchasing decision. You can vary your ads somewhat but keep them consistent in visuals and message.

Your customers have busy lives so keeping your name visible will remind them that they want to stop by and see what you have for sale – instead of picking up a loaf at the grocery store. You'll also remind customers that they haven't stopped in for a while to see what's new.

Chapter 9

Public Relations: How to Get Customers in the Door with Little Cost

What Is Public Relations?

Public relations are really the sum of its many definitions. It's the message a person, company or organization sends to the public. It's a planned effort to build positive opinions about your business through actions, and communications about those actions. In short, it's any contact your organization has with another human being, and the resulting opinion. This opinion may or may not be accurate, but it comes from everything the public reads, sees, hears and thinks about you. Effective PR has been described as becoming a positive member of your community (and getting credit for it). Good PR sends a positive message to the public about your establishment.

PR should be part of your overall marketing communications program. This includes advertising, internal communications, and sales promotion. Speeches, contests, promotions, personal appearances, and publicity are parts of PR, but really, the results generated from all of these parts—including acquiring unpaid-for media space and time—are PR. It's who the public thinks you are and the nurturing of that opinion in a positive way.

What PR Does (And Doesn't Do) For You

If done well, PR distinguishes you from the pack in the eyes of your customers. It leaves them with a favorable impression of you and great tidbits of information to pass on to their friends about your establishment. It makes you newsworthy in a great way and can help save your reputation and standing in your community during an emergency.

Good PR improves sales by creating an environment in which people choose to spend their time and money. As said before, PR is getting credit for being an upstanding member of your community. If you are not, PR can't make you look like you are. PR accentuates the positive and creates lasting value by highlighting what makes your establishment special. PR cannot create lasting value if none is there to begin with. What it can do is communicate existing value effectively, so it lives in your customers' minds.

Good PR can make a good story great, and a bad story less bad. However, PR is not just the public's opinion of your business; it's also the physical state of your establishment. People aren't just interacting with your staff; they're interacting with your facility. If the media are reporting on something wonderful that happened at your bakery, but the place is in a state of disrepair, what are you communicating about your establishment?

The key to implementing an effective PR campaign is determining what your business's image is, what you want it to be, and how best you can create that image in the eyes of the public. You need to clearly define your objectives and create a plan that will implement them. PR is not a way to gloss over a tarnished image or to keep the press at a safe distance; it's an organized and ongoing campaign to accentuate the positives of who you truly are.

PR is Different from Advertising

PR is not advertising; PR uses advertising as one of its tools. A good PR campaign is almost always coordinated with advertising, but PR is not paid-for time and space. In advertising, clients pay the media to carry a message and the client has complete control over this message. With PR, the media receives no money. Because of this, your story about the famous movie star dropping by your bakery may end up on the 5 o'clock news, in the paper or nowhere at all. The success of a PR story often depends on how timely it is or whether an editor feels it's worth reporting on. Furthermore, only a portion of your intended message may be used. The media may not even use your bakery's name. Because they are choosing to write about your topic, and you've basically given them only a potential *idea* for a story, the story could end up in a very different form than you initially presented or hoped.

PR doesn't give you the control you have with advertising. However, when done well, it garners positive attention for your establishment, is hugely cost-effective, and is *more credible than advertising*. This is because the public is getting its information from a third party – not directly from a business. Customers assume advertising to be self-serving; but a positive message delivered by a third party is perceived to be authentic and trustworthy. *Third-party messages are infinitely more persuasive than advertising.*

The Marriage of PR and Marketing

Public relations are one of the crucial aspects of a successful marketing plan. When management is communicating effectively with guests, employees, and community leaders, it is implementing an effective marketing plan. Fundamentally, all marketing is integrated. Consumers don't distinguish between one message from your business and another—all the messages are yours. In that light, because it's your job to communicate as well as possible, understanding that all your marketing is integrated allows you to focus on an overall approach to building good PR.

Launching a PR Campaign

In a small bakery, the manager may be solely responsible for public relations. In a larger establishment, the director of marketing or sales often plays this role. Having a single person designated as media liaison makes it simple for the press to get their questions answered and makes it much easier for you to control the flow of information to them. This back-and-forth is a critical element in your PR campaign. Once this liaison is determined, notify your staff. Advise them not to talk with the press, but to refer all media inquiries to your liaison.

Start Your PR Campaign Now

Public relations campaigns for a new business or new ownership of an existing business should start BEFORE your doors open. Start by creating buzz within the community.

- Post "coming soon" signs.
- Look for complementary businesses (vendors, neighbors, banks, peers) to share the PR costs.
- Contact local media for business-oriented PR (business article in newspaper).

In launching your campaign, it's important to remember that you will be competing with professionals for a very limited amount of airtime and/or editorial space. Reading newspapers and trying to determine which pieces were inspired by PR people – and what about them made editors choose them – is a good discipline. In addition, many community colleges offer courses in small business public relations. The more expertise you have, the more effective your campaign will be.

If your establishment is part of a franchise, PR assistance may be available from the franchisor. If you manage an independent property, PR help may be available from your local Better Business Bureau, Chamber of Commerce or Convention/Visitors' Bureau.

How to Apply Your PR Plan

Once you have established the objectives of your PR campaign and integrated them into your marketing plan, it is time to execute. These questions can help you do just that:

- What's the right medium for this strategy?
- Who are the key contacts?
- How strong are the necessary personal relationships required for this plan? Do any need to be established or reestablished?
- Is this plan thorough? Have we considered all the risks?
- Are we prepared to deliver a press kit to selected media contacts?
    - This press kit is an essential part of your plan. It contains background information, newsworthy facts, contacts, phone numbers—all the pertinent information that will inform the media and direct them to you.

The press may not use one word of your materials, but there is a much greater likelihood they'll describe you the way you want them to if you've given them the resources to do just that.

When providing the press with information (spoken or written) be:

- **Honest.** The media want credible, honest material and relationships. Your message should be genuine and factual. This doesn't mean you have to reveal confidential data; it just means that your materials should be thorough and truthful.

- **Upfront.** Don't lie, dodge or cover up. If you don't have every answer to a question—and you might not—don't say "no comment," or "that information is unavailable." Simply respond that you don't have that information, but will provide it as soon as humanly possible. Then provide it as soon as humanly possible.

- **Factual.** Give the facts and follow up. If you supply the media with a printed handout of key facts, it greatly lessens the chances of your being misquoted. Make a concentrated effort to follow up and go over information with the media. Again, if you don't have a requested piece of information, get it and follow up with a note and/or call to make sure the correct data reaches the media.

- **Concise.** The media is more apt to burn you for what you say, not what you don't. Be deliberate about providing the facts without editorializing, exaggerating or pulling things out of thin air.

- **Professional.** If you follow the above steps you're on your way to building a strong and lasting relationship with the press. These relationships can sour instantly if you are reactionary, hostile, aloof, hypersensitive or argumentative in any way. No matter what you think of an interviewer, treat him or her with respect, courtesy, and professionalism. Causing negative reactions from the press will deny you print space and airtime.

How you interact with the press is crucial, but it's only half the process. The content of what you communicate to them—having a clear and deliberate focus about how you are going to tell your story—is the other side of press relations. The following list will help you identify your purpose and communicate it effectively to the press:

- **Identify your purpose.** Why do you want public exposure? What are you specifically trying to draw attention to? Are you selling your bakery's expansion? Then don't go on about its famous muffins. Be sure you are conveying your purpose.

- **Identify your target audience.** Who are you targeting? Prospective customers? Your employees? The local business community? Civic leaders? Lay out whom you want to reach, and then determine who in the media will speak to them most effectively.

- **Look at it from the media's viewpoint.** Why would this be interesting to the media? Figure out how your interests can be packaged in a way that directly matches the press's interests. Make your story one they want to print—i.e., one that will help them sell papers, gain listeners, etc.

- **Customize your materials.** Once you have identified your purpose, who your target is and the media's angle, tailor your materials to include all three. Give the press everything they need to tell the story—photos, copy, etc.—and be sure it's in exactly the style and mediums they're using.

- **Know where to send your materials.** Is your story a business story or a feature story? Do you know the difference? A business story goes to your newspaper's business section editor. Feature stories go to the appropriate editor: food, community, lifestyle, etc. It's a very good idea to cultivate relationships with these editors beforehand so that when the time arises, they are thinking well of you and would like to help.

- **Make their jobs easy.** Do not ask the media for the ground rules for getting press and building relationships—learn these on your own and then meet them. Spending valuable time and resources building a relationship with a reporter, only to then submit materials at the last minute or give them insufficient or inaccurate information, burns bridges quickly. Do as much of their work for them as possible: give them something that is ready to go, answers all their questions and is interesting. This is the difference between staying in the middle and rising to the top of a busy person's in-box. Also, be available immediately to answer questions. If a reporter calls and you aren't there or don't return the call immediately, your great story—prepared at considerable expense—may end up in the trash.

Before you begin your media campaign, you should get to know the media as much as possible. This may mean inviting them – one at a time – for a brief tour of your establishment and, perhaps, an afternoon tea. This gives them a connection to you and your business and begins to build a relationship.

These visits are NOT the time to sell them on doing a story on you. It's a time for you to get to know each other and to build a relationship. If the reporters trust you, they will help you, and vice versa. They need article ideas as much as you need press, and getting to know them will give you insight into how you can help them do their job.

A secondary goal is that you will be establishing yourself as a bakery/pastry expert. When they need a quote or some advice – you'll be the one they call!

Once your friends in the media trust you won't be barraging them with endless story ideas, you can begin your media campaign. It is important to remember that having a positive rapport with a reporter doesn't mean he'll do a story on you. Your relationship with the reporter will help get a newsworthy story printed, but you won't get a boring story to press just because the reporter likes you. ***Your story needs to be newsworthy on its own.*** In addition, reporters are always working against time. The more you can give them pertinent, accurate, concise information, the better your chances of getting their attention.

If you've built a respectful relationship with the media, a reporter who gets a story from an interview or news conference at your establishment will mention your place in her story. These are the "freebies" that come from developing strong relationships with the media and learning to think in their terms.

Many businesses go one step further and give their media contacts news releases that are written in journalistic style. A news release describes the newsworthy development in your bakery in a ready-to-print article. Editors can then change it or print it as is. These can be immensely valuable for getting your message out there.

If writing journalistic articles is beyond your reach or budget, tip sheets can be very effective in getting your story across. A tip sheet gets the message to the media by simply outlining the *who, what, when, where, why* and *how* of your story. It's basically an outline of the story the reporter will then write. Tip sheets give the spine of the story and, because they are so concise, often get more attention from busy editors.

Here are a few more tips on how to work effectively with the media:

- Earn a reputation for dealing with the facts and nothing else.
- Never ask to review a reporter's article before publication.
- Never ask after a visit or an interview if an article will appear.
- Follow up by phone to see if your fact sheet or press release has arrived, if the reporter is interested and if she needs anything else.
- Provide requested information—photos, plans, etc.—ASAP.

Creating Your Press Kit

It may be a good idea to hire a part-time PR consultant, former reporter or editor who can help you present your materials to the press. If this is beyond your budgetary limits, the following is a list of essentials for creating your own press kit.

- **Fact Sheet** – One of the most helpful items of media information, the fact sheet does most of the reporter's research for her. It also shortens the length of interviews by answering questions in advance. It should describe your business and what you are trying to get press for. At a glance, it tells where you are located, when you opened, your food offerings, specialties, and number of employees. It should also specify the types of facilities you have, such as historical information on the building and what type of atmosphere your bakery has (family-friendly or upscale singles).

- **Staff Biographies** – You will need to write biographies for all of your key executives. These list work experience, education, professional memberships, honors, and awards. Include information on ties to the community and other human-interest angles.

- **Good Photography** – Do not take chances with an amateur photographer. Space is *very* limited in the print media, and editors go through *thousands* of photographs to choose just a few. This is true even for local editors. Don't give them any reason to ignore your pictures. Have them taken by a pro. Ask for references and check them thoroughly. When the photos are done, write an explanatory caption for each picture in your collection. This gives editors an easy understanding of what they're looking at. Then, before sending photos to the media, be sure you find out whether they prefer black and white, slides, transparencies, etc., and send them in the desired format.

- **Press Kit Folder** – Put all of these materials into a single folder with your bakery's name and logo on the cover. You might also include brochures, a brief on your involvement with local charities, etc. Don't overstuff it, but give the press a solid idea of what distinguishes you from the competition. **Do not include press clippings! You'll diminish your newsworthiness.** No respectable editor/reporter want to rehash something that has already been done. The rare exception to this might be that your bakery was featured on the cover of a national food magazine and your news angle is *"local bakery receives national recognition."*

What's News?

Once you have identified your target media and begun your media relations program, you need to learn what makes news. To do this, pick up the paper, and turn on the TV. The media is looking for the strange, volatile, controversial, and unusual. It's not newsworthy that you run a nice bakery that provides great food at a reasonable price. It's newsworthy when a customer gets food poisoning at your bakery, or when a group's convention reservations are cancelled. This is not the type of news you want to make, but it's news. Obviously, you want to be making great news. One of the foundations of this is taking steps to avoid negative articles: making sure your food safety measures and sanitation ratings are excellent, your staff treats guests courteously, etc.

Once you've taken these steps, you are ready to generate positive stories in the media. How? Well, what do editors find newsworthy? Here is a list of basic newsworthiness criteria:

- Is it local?
- Is it timely?
- Is it unique, unusual, strange?
- Does it involve and affect people?
- Will it provoke human emotion?

Think in terms of what it is that sets your establishment apart from the competition and what is newsworthy about those qualities. When this is done, again, target your media. When you have a story, be smart about who would be interested in writing about it and whose audience would love to read about it. Here is a short list of possibly newsworthy ideas:

- A new manager or pastry chef.
- Visits by well-known politicians, entertainers, authors or local heroes.
- Private parties, conventions or meetings of unique organizations: antique car enthusiasts, baseball card collectors, scientific organizations, etc.
- A new menu item or specialty baked goods. (e.g. egg free, low carb)
- Hosting a charitable event, sponsoring a school activity, feeding the hungry.
- Reduced rates, special menus, promotions, weekend specials.
- Personal stories about the staff: an employee who returned a doctor's medical bag, helped a patron stop choking, returned a tip that was too big, etc.

Choosing Your PR Contacts

The first goal in building strong media relations is to determine who your target media are. Discover which news media outlets reach your targeted audience and the means they use to carry their messages. Your target media will change according to the type of message you wish to send and the type of audience you wish to reach.

Once you know who your target is, the next task is to build a media list. This list includes names of appropriate editors, reporters, news directors, assignment editors, media outlets, addresses and contact numbers along with demographic information. If you already have an advertising agency, they can either assist you with PR or refer you to a skilled consultant to handle the task. The agency can also gather and interpret PR statistical data for you. However, most small businesses won't have the budget to hire a PR firm so the task is left to management.

If you want to mail fact sheets, press releases, press kits, etc., you can hire a company that sells media mailing lists, and you can pay them or another firm to do your mailing for you. If that is beyond your budget, you'll need to spend some research time creating your own list. If you are handing your own publicity, you'll have to do some Web research and/or telephone work to find the right person within each organization.

Start by calling the editorial department of a newspaper or a newsroom to get the contact numbers of the people you seek to reach. Your assigned advertising rep for the TV, radio, newspaper or magazine may be able to direct you to appropriate editorial staff people.

Your first call should be to the food editor of your local paper. A mention in a column or recipe for a home baker in an article are valuable ways to create great "buzz" about your business. The business editor wants to know about your business skills, sales techniques, expansion efforts or technical advances within your business. The lifestyle editor wants to know about your leadership and charitable activities. The community/neighbor editor is interested in the local angle.

Magazines, radio and TV stations will have similar editorial managers each with their own responsibilities and agendas. Your job is to find these people and to tailor your news information to their needs.

In addition, you may want to target national media, as well as specialized trade and business publications. Online publications and websites dealing with food-related topics (trade and consumer) can also be a way to get your name out to the public.

Developing Allies

During your campaign, it's also important that you search for allies. Allies are businesses and organizations that have similar goals to yours. Your state's Tourism/Travel Promotion Office can be a great resource for this. This office is working year-round to bring business and leisure guests to your state. These, of course, are your prospective customers. Your state's travel promotion officials will be happy to give you advice on how to tie in with their advertising, PR ,and other promotional programs.

Most states also have a business/economic development department that will be happy to help you, since their goal is to create new business in your state. Their mailing list will keep you informed of planned promotions. When meeting with state officials it's a good idea to volunteer to assist their promotional and PR programs. Doing this gets you "in the loop" and, often, ahead of your competition, because you'll know about the programs your state is developing. There are a number of national travel industry organizations that work privately to generate travel in the United States. Hospitality and food service associations can also prove to be valuable allies, since they either have PR people on staff or use national PR agencies.

Locally, your Chamber of Commerce, convention center or tourism department may organize trips to your area. These are trips for travel writers and travel agents that showcase the attributes of your area. Let the organization arranging the introduction trip know that you're willing to offer free refreshments or meals to the visiting journalists and travel agents. If you are selected, make sure time is allotted for a guided tour of your property, led by your most knowledgeable manager or salesperson. Present each guest with a press kit. Also, mail press kits to the agents after the tour, since most of them prefer to travel light but accumulate tons of literature and souvenirs on their trips. Making a good impression with travel agents and writers gives you a chance for exposure beyond your own community.

When these agents and writers do visit, make sure that your establishment is in tip-top shape. Your visitors will probably be visiting numerous other hospitality and food service establishments and you want to stand out in every positive way. Only the most memorable businesses will be on their "recommended" list and you want to be one of them.

Charity for PR

As a good citizen, supporting your local charities is probably something you already do. As a businessperson, you can also use these to enhance your image within the community – a win-win situation for sure. Marketers know that purchasing decisions are based upon emotions. By putting a human face on your business you'll create a positive image within your community.

There are a variety of non-profit organizations that you might consider supporting. Your first concern should be selecting a worthy charity. Look for a charity that is well regarded within the community, fiscally wise, and worthy of your time and money. The secondary aspect is the PR value of the support. Look for charities that give you a chance to "show off" your baked goods first. Not only will you be helping to raise money, you'll also be introducing your bakery to the event attendees. There is nothing wrong with clarifying how your support will be recognized and providing the charity with appropriate logos in your press kit. Fundraising experts will understand and won't make you feel like you are "buying" advertising.

A wonderful aspect of charity PR is that you can have a lot of fun. Coach the Little League Team you sponsor. Operate the fundraising booth. Serve up a slice of pie. You'll get out of the kitchen and discover the benefits of helping others go well beyond public relations!

Special Events

Special events can be very effective in generating publicity and community interest. You may be opening a new property or celebrating a renovation or an anniversary. Any such occasions are opportunities to plan a special event that will support or improve your PR program. There are usually two kinds of special events: one-time and ongoing. Obviously, you're not going to have a groundbreaking ceremony annually, but you might have a famous Fourth of July party every year.

The key question to ask when designing a special event is "Why?" Clearly defining your objectives before you start is crucial. Is your goal to improve community opinion of your business? To present yourself as a good employer? To show off a renovation?

If you are planning an anniversary celebration, research what events were going on in your community when you opened: Was there a huge fire? Had the civil war just ended? Did Dwight Eisenhower speak at the local college? Once you have this information, send it to the press. They will see your event as part of the historical landscape, as opposed to a commercial endeavor that benefits only you, and they'll appreciate your community focus.

Special events require preparation to ensure everything is ready when the spotlight of attention is turned on you. Be certain the day you have chosen does not conflict with another, potentially competing event, or fall on an inappropriate holiday. With a groundbreaking or opening of a new property, you should invite the developer, architect, interior designer, civic officials—all the pertinent folks—and the media. You should prepare brief remarks and ask the architect to comment on the property. In your remarks, remind your listeners that the addition of your business does not boost school taxes or increase the need for police and fire protection; it adds new jobs and new tax revenues.

If you are celebrating an opening, tours of the property are a must and should be led by your most personable employees. Refreshments should be served and plenty of food samples available. Whatever your occasion, you should provide press kits to the attending media and mail them to all media that were invited. Souvenirs are a good idea—they can be simple or elaborate, but should always be creative, fun and useful to your guests.

Talking to Your Community

Your bakery success depends upon your community. Bakeries that are not accepted by their local communities disappear. It's as simple as that. In addition, you won't find a prosperous bakery in a depressed area. Your community and you are one and the same, and it's crucial to remember this as you design your PR program. Your bakery can't be successful unless the local community accepts it.

So what does that mean to your bakery? It doesn't simply mean that you should help support good causes. It means your business needs to be a leader in its community. In practice, this means building bridges between your company and your community to maintain and foster your environment in a way that benefits both you and the community. Basically, your goal is to make your immediate world a better place in which everybody can thrive.

The following are a few ideas that can be part of an effective community relations program:

- Fill a community need—create something that wasn't there before.
- Remove something that causes a community problem.
- Include "have-nots" in something that usually excludes them.
- Share your space, equipment or expertise.

- Offer tutoring, or otherwise mobilize your workforce as a helping hand.
- Promote your community elsewhere.

Being a good citizen is, of course, crucial, but you also need to convince your community of the value of your business as a business. Most businesses provide jobs and pay taxes in their communities. Because bakeries are labor-intensive businesses, they probably provide employment for more residents of your community than any other business based on per dollar of income.

Your bakery can become a community hub. Your dining area can become a meeting place and a center for important social/economic functions. Decisions that affect the future of your local economy can happen over a donut and cup of coffee in your bakery.

These are real benefits, and they should be integrated into the message you send by being a good citizen. Designing this message is a straightforward but remarkably effective process:

- List the things your establishment brings to the community: jobs, taxes, well-maintained architecture, etc.
- List what your business receives from its community: employees, fire and police protection, trash removal, utilities, etc.
- List your business's complaints about your community: high taxes, air pollution, noise pollution, narrow roads, etc.

Once you have outlined these items, look for ways your business can lead the way in improving what doesn't work. As you do this, consult with your local chamber of commerce or visitors' bureau. They may be able to integrate you into existing community betterment programs aimed at your objectives.

If done well, your community relations program will create positive opinions in your community. In turn, this will cause local residents to recommend you to their friends, encourage people to apply for jobs, and may encourage suppliers to seek to do business with you. In addition, if there is an emergency at your establishment, having a positive standing in the community will enable your employees and business to be treated fairly.

An effective community relations program is a win-win situation because it gives you the opportunity to be a deep and abiding member of your community—improving the quality of life and opportunities around you—and, at the same time, contributes significantly to your bottom line.

Employee Relations Is Also Public Relations

One of the most important "publics" that your public relations program should focus on is your staff. Happy and productive employees are vital to good customer service and they share their feelings outside of work – that's good PR all the way around!

Customers want to be taken care of, and they judge a business as much on the quality of the service as the product. Basically, if a member of your staff is grumpy or tired, that's bad PR. Therefore, employee relations should be a main focus of your PR campaign. In order to do this you must have well-trained staff that understand the technical ins and outs of their jobs and also believe in your organization's mission. Your employees need to know the high level of service your customers expect, and they need to be empowered to deliver it. A staff that does this on an ongoing basis is one that generates repeat business through word-of-mouth referrals and good PR.

Keeping your employees informed is a key way of making them feel involved and building positive feelings between staff and management. The following is a list of things to communicate to your staff:

- How your business is doing and what you're planning.
- How the competition is doing and what you're planning.
- What community issues you're concerned about and taking a role in.
- Recent personnel changes.
- Available training and job openings.
- Staff weddings, birthdays, significant accomplishments or happenings.
- ASK, LISTEN and ACT. Don't just tell them – show them how important they are to your business.

Communicating this information gives employees the sense that you care and creates a unified work atmosphere where great service becomes a group responsibility. It also shows that you recognize the difference they make to your bottom line and that you're paying attention.

Opening the lines of communication between management and staff is the next step. No one knows the intimate ins and outs of your business like your staff. If they care about your business and know your ears are open, they can be your biggest resource in suggesting improvements and letting you know what's really going on. One-on-one meetings with supervisors, group meetings, employee newsletters, orientation/review sessions, and training meetings are all effective ways to open the channels of conversation between you and your employees. These sessions let them know you care and encourage them make the biggest difference they can.

An ongoing employee appreciation program is a good idea. Create a structure that is a part of your daily operation: a large bulletin board in a high-traffic area or a monthly party where awards and prizes are given (cash, great parking spaces, etc.). Give employees something to wear (e.g., a recognition pin) that signifies acknowledgement of the services they provide. Be creative, and find something that effectively and continuously supports the goals of both your employee relations program and your overall PR plan.

Planning For The Unforeseen

When you are knee-deep in an emergency is not the time to develop a crisis plan, *you must prepare now*. If you have a strategy developed in advance, then when something bad does happen, you assure the most accurate, objective media coverage of the event. It's important that all your employees are aware of this plan and that they are reminded of it regularly. Since your employees generate a huge amount of your PR, it's crucial for them to know how to act and what to say—and not say—during a crisis.

When an employee is injured or killed in your establishment or a guest suffers from food poisoning, the public assumes you're guilty. Whether or not you're even mildly at fault, people assume you are. Therefore, how you handle public relations during this time means the difference between a temporary loss of public support or the permanent loss of a great deal of your business.

Having built strong media relations pays off during an emergency. The person who will be the media liaison during an emergency should be building and nurturing good media relations now, in case anything does happen. With proper PR, you can shift the public from viewing you as incompetent to having more faith than ever in your establishment. Public opinion depends on how effectively you manage information and how well you get your story across.

Chapter 10

Choosing Your Product Mix

• •

Your bakery may be a wholesale operation, a retail store, a deli, a diner, a candy shop, a tourist attraction, a wedding supply shop or a combination of any of these. What you *make and what you purchase for resale* are your product mix and/or menu. Your job is to select items that your customers **need** and **want**; that you can buy or make at a **reasonable price**; and that you can be **profitable** selling. In this chapter, we'll divide the topic into *food, beverage* and *non-food items* and discuss some of the choices you might offer custom-

ers. In Chapter 11, we'll discuss recipe development and production.

Food

Your establishment will, of course, be selling a variety of baked goods – sweet and savory. Whether you are selling wholesale, retail or both, you'll need to offer the breads, cakes, cookies, and desserts that your customers want. How do you know what they want? In researching your market, dreaming up your ideal customer and writing your business plan, you should have discovered some needs that you can fill.

Food Trends

Safety, health, indulgence, and convenience are all growing trends in food. You'll find your customers are interested in one of these – and sometimes all four – even if they are contradictory! The key to tapping into food trends is to ***tell them about it***. Tell your customers all about your ingredients and how they are good for them (physically and emotionally), how luxurious your food is and how easy it is buy and use. Include descriptive copy and plenty of adjectives!

> **Safety** – allergen-free, kosher, chemical-free, natural ingredients, organically grown.

Unlike "health" below, safety goes beyond the personal well-being of your customers and includes a desire to protect the earth by using products grown and produced without harming the environment.

> **Health** – vitamin-rich, naturally rich in nutrients and fiber, filled with antioxidants, prevents disease or illness.

Healthy food choices may be based upon scientific evidence or the latest diet fad. Read the Health section of your local paper to keep in touch with health-related food trends. As a small business, you have the flexibility to take advantages of health-related trends. Allergy issues are less affected by trends. By learning about your customers, you can provide special allergen-free products (no eggs, dairy-free, no nuts, etc.) Vegetarian products and kosher foods bridge multiple categories for consumers who are accustomed to paying extra for peace of mind.

> **Indulgence** – rich, creamy, buttery, laden with expensive ingredients.

You'll have customers who will only buy the most indulgent of desserts and you'll have others that will only splurge for special occasions. You'll even have those who want healthy offerings Monday through Friday and the diet-busters during the weekend. These customers feel that they deserve it – a food reward.

> **Convenience** – ready-to-heat packaging, precut, enhancements.

Busy lives means less time to prepare meals. Creating food that is quick to serve, yet has a homemade feel appeals to "guilty" parents. Handy serving sizes and resealable packaging speed things up, too. Don't forget to offer enhancements such as flavored butters or cream cheese, whipped cream, and nut sprinklers along with serving enhancements such as birthday candles and decorative plates. Make your products the perfect way to transform an ordinary meal into extraordinary with no work!

Your Niche

Most bakeries offer a wide variety of baked goods. However by concentrating on a specialty product or two, you'll develop a reputation and hopefully fill an underserved marketing. Look carefully at your competition – examine what they are offering and what they aren't. Look for "weak" spots in their quality, presentation and selection.

Bread

There are literally thousands of types of breads – from tortillas to scones to hamburger buns to sourdough to cinnamon rolls to English muffins to French bread. Your bakery may focus on small production hand-kneaded artisan breads or wholesale production of a dozen varieties of bagels.

As you review your potential choices, you'll also need to consider:

- **From-scratch vs. mix or partially prepared** – Will you be purchasing raw ingredients, bulk mixes or ready-to-bake items? To choose, consider:
  o   Costs (labor, raw materials, overhead).
  o   Quality of finished product
  o   Ability to produce volume in required time.

- **Equipment Needs** – Will you need costly specialized equipment? Will you need to increase your equipment's capacity?

- **Facility Requirements** – Will your plant have sufficient space for production? Will your anticipated volume be feasible in the space allotted? Do you have the proper electrical set-up?

- **Employee Needs** – Will you need to hire additional staff? Will additional training be required (on-site or formal school)?

Going Natural

The words "whole wheat", "natural", and "organic" have been tossed around for years in food manufacturers packaging and marketing. Although the U.S. Food and Drug Administration passed National Organic Standards (info at **www.ams.usda.gov/nop**), most "healthy" terms now have official federal definitions and standards.

- To learn more about organic foods, find retail/manufacturer training materials and order promotional aids for organic products, visit the Organic Alliance (**www.organic.org**).

- For information on retail compliance of the NOS, visit this USDA Web page at **www.ams.usda.gov/nop/ProdHandlers/ RetailFoodEstablishments.html**

- You'll find a helpful consumer's guide at Eco-Labels (**www.eco-labels.org**).

You'll find more information on product labeling in Chapter 13.

Desserts

Feed your customers' sweet tooth to feed your bottom line. Industry statistics show that cakes typically generate 30 percent of sales for an average full-line bakery. Learn tips and trends for generating positive results from your dessert cake programs as well as other bakery components in the bakery-focused section of SuperMarket News (**http://supermarketnews.com/product-categories/bakery**).

A variety of dessert items can be added to your product mix that can be sold whole (like tarts or cheesecakes) or sold in individual servings (mousse cakes or cobblers). Individual servings are becoming extremely popular as the number of single diners continues to increase across the U.S. Review your customer demographics to see what trends are happening in your neighborhood.

Specialty Bakers

Your passion for baking might lead you to be a specialty baker – where you focus on a single product type such as bagels, donuts or wedding cakes. Just as you would explore your niche – centering your attention (in production, service and marketing) on a "specialty" item is an excellent way to build a brand name and a reputation for expertise. However, the downside can be that you discover your choice is too limiting financially. It can be difficult to reposition your business with customers who only think of you for a specific item.

Other Food Items

Beyond your own baked goods, you might consider offering ice cream by the scoop, candy or other impulse food items. Your choices should appeal to your customers and take into account your foot traffic and tourist appeal. Other foods like a bag of your special coffee blend, an imported balsamic vinegar, virgin olive oil or flavored butters would be excellent enhancements for your baked goods – sweet or savory!

Serving Meals

The coffee craze has transformed many retail bakeries into places to linger over a cup and a sweet roll. Deli/bakery combinations that offer a limited lunch menu featuring your own baked goods can be a great profit center. You may find serving prepared snacks, sandwiches and a few entrees a great way to attract a new crowd. Chapter 11 discusses developing a menu for prepared take-out or dine-in service.

Beverages

The first non-bread/dessert products most people will add to a bakery menu are beverages such as:

- Coffee - plain, decaf and flavored
- Coffee – iced or hot coffee-based drinks
- Tea - iced or hot
- Milk - non-fat, 2% and chocolate
- Juice (orange and other local favorites)
- Trendy drinks - juice-based, tea-based, herbal or energy drinks
- Soda - Coke® or Pepsi® products

Coffee

You could buy all your own equipment and pay to have someone install and maintain it. This would enable you to purchase coffee from any company at reduced prices. However, a large capital outlay would be necessary. Since there is no great advantage to doing this, it is recommended that you use the coffee distributor contract method.

There are many different coffee blends available. Coffee is an extremely important part of any dine-in or take-out bakery service. Commuters look forward to a steaming cup of coffee, cappuccino or latte with their morning bagel or Danish. Invest in the finest and most popular blends available. Try the different blends under consideration in a blind tasting.

Specialty Coffee Machines

Coffee is no longer a simple mug of decaf or regular joe. The growth of specialty coffee drinks means your customers will be asking for a grande vanilla latte or a raspberry mocha. Espresso machines can run into the thousands of dollars and require a skilled barista to operate. Your coffee vendor can help you assess your needs based upon customer demographics, volume and budget. Espresso systems can be expensive to operate and maintain so explore your options before signing a contract.

For more information on espresso machines, visit Espresso Business at **www.espressobusiness.com** and Whole Latte Love at **www.wholelattelove.com**.

Selling Something Other than Food

The *good news* is that people are willing to purchase non-food items and services from independent bakeries. This translates into potential new profit centers. The *bad news* is that you could be risking your core business. If customers stop thinking of you for your food or you spend too much time and energy looking, you may jeopardize your bakery's financial health.

Complementary Goods

The most obvious choices for non-food items would be goods that enhance your customer's use of your bakery items such as candles for birthday cakes, party ware and insulated coffee mugs.

Last-minute gift items and greeting cards are perfect for someone who rushes in to pick up a birthday cake. Souvenirs and branded-items (mugs with Betty's Bakery) may be good choices. Floor space can be a very valuable commodity; so don't crowd customers with knickknacks.

Chapter 11

Profitable Menu Planning — How to Get Maximum Results

For those who wish to go beyond retail baked good sales, planning a successful and profitable menu for a deli or café will require an understanding of your competition, your customers and your facility restrictions. Prepared food sold for take-out or dine-in should be considered an additional profit center. Your primary responsibility is to focus on your core business – your baked goods. Everything else you sell or do should be a profit-making enhancement to your business.

Going Beyond Bread

Before expanding your service and products, review your business pluses and minuses.

- Assess the strengths of your staff. Are they capable of handling additional duties?
- Do you have time to develop and launch a new facet to your business?
- Do you have ample room for the venture or the ability to expand your facility?

- Are you thinking about doing this to grow your business or to save a failing one?
- What resources can you draw upon to ensure the success of a new profit center?

Your Goal

Before figuring out your menu, you need to decide what you are trying to accomplish.

- Do you want to feature *existing* products in new ways? Selling sandwiches means you sell more bread.
- Do you want to add *new* beverages or food items?
- Do you want to make it easier for people to pick up orders or to eat your food?
- Do you want customers to think of you as a comfortable place to linger or dine?
- Do you want people to have more reasons to come in at other times of the day?

It would be impractical in these pages to list specific examples of potential menu items. There are many excellent cookbooks that describe menu and recipe ideas in detail. The following sections will illustrate a basic outline, from which you can plan your own exclusive menu. Each of the procedures described play an integral part in developing your cost control system. Adapt the procedures that suit your own particular needs. However, the purpose of creating menu items is to increase profits so you must watch your development cost along with ongoing labors.

Menu Style

The things that may influence whether or not you offer a limited menu include kitchen size and labor cost control. Menus with more options do have a broader appeal; however, by keeping production simple, it will be easier to control costs and provide a quality product – without sacrificing your bakery's service.

A fine-tuned, but limited menu can be tailored for your bakery to offer three meals a day or serve just the breakfast crowd. By preparing and combining the same ingredients in different ways, your can offer more choices but still control inventory and costs.

So, how many menu items should you offer? You want to provide the customer with variety, but not at the expense of your ability to control inventory and cost, nor by overtaxing your production or service staff. Research has shown that 60–75 percent of menu items sold are the same 8–12 items, regardless of the number of choices offered. Keeping this in mind, it is probably wise to offer somewhere between 18 and 24 options.

Formatting Your Menu

After you have defined you establishment's goals and determined the style of your menu, you must decide what items will go on your menu. This can be done in a five-step process:

1. You must decide what menu groups you will offer. Groups are snacks, breakfast items, entrees, soups, etc.

2. Decide what categories to offer within these groups – kosher, health conscious, meat eater, and vegetarian. One caveat – if you are offering kosher or vegetarian foods make certain that all ingredients comply with the strictest definition of the terms. Don't forget to provide ingredient lists.

3. Decide how many items you will offer in each category. After deciding the groups and categories that will go on your menu, you need to decide on the preparation specifics of the item. Will you be serving ground, cubed, solid, or roast? Baked, grilled, broiled, fried or toasted?

4. Finally you must decide on the recipes (even if this is just step-by-step instructions on how to prepare a toasted bagel with lox). This will help you develop food inventory, purchase needed equipment or prep tools, develop workflow patterns and establish training procedures.

While this may sound like an onerous task, by keeping these five steps in mind you will be able to maintain variety in your menu and control cost factors.

Developing the Menu Selections

All menu items selected must fit into the physical workings of the bakery. If possible, this menu should be finalized prior to designing, selecting equipment for and laying out the kitchen. However, if you are purchasing an existing bakery operation or adding to your current product mix – then you'll need to develop menu items based upon your physical ability to produce quality food quickly and efficiently.

The design and layout of the kitchen and work areas must meet the needs of the me nu. If they don't, the entire operation will become slow, disorganized, and inefficient. Inefficiency lowers employee morale and your bakery's profits

Just as the kitchen must meet the demands of the menu, the personnel employed to prepare the menu items must be a fit. Careful consideration must be given to the number and type of employees needed.

- Is the menu simple enough for inexperienced workers to prepare or are the skills of a professional, more experienced cook needed?
- Will the food be prepared ahead of time or upon receipt of the order?
- When will these employees be needed, and for how long?
- Will there be enough room in the kitchen for everyone to work at the same time? Who will supervise them?

Planning the bakery menu is a lot more than merely selecting menu items that are enjoyed and demanded by the bakery's clientele. Menu development includes purchasing equipment, hiring personnel and sourcing food products. Your goal is for the item to be affordable and in demand by the public. Successful, growing bakeries have accomplished this blending—the ones that failed, didn't.

Having separate preparation and cooking staffs is the most efficient method for producing a large number of consistent products at the least expense. Since this is the method used in most food service establishments, it will be the one described and referred to in this book.

The Major Points To Consider When Selecting Menu Items:

1. The menu item must be of superior quality.
2. Menu items should be creative and feature your baked goods. Try to include items not readily available in other bakeries.
3. The raw materials used in preparing the item must be readily available year-round at relatively stable prices.
4. The raw materials used in preparing the menu item must be easily portioned by weight.
5. The menu item must be affordable and demanded by your customers.
6. The menu item must within the skills of your preparation and cooking staff.
7. All menu items must have consistent cooking results.
8. All menu items must have a long shelf life. Some food items prepared ahead of time may not be sold for as long as 36 hours.
9. All menu items must have similar cooking times as any entree requiring a longer cooking time will not be completed when the other orders are ready to serve.

10. All menu items must be delivered quickly as your customers will expect this in an expanded bakery environment.

11. The storage facilities must accommodate the raw materials used in preparing the menu items.

Limiting the Menu

Begin to develop the menu by compiling those recipes and ideas which meet the requirements set forth in the previous section. Consider only the items which are compatible with the bakery's atmosphere, decor and anticipated clientele. Based upon these guidelines, you should have little trouble compiling a considerable list of acceptable choices. The trick is to limit the menu to only those items for which the kitchen is equipped and organized and that the staff can easily execute – while still allowing for an interesting menu with plenty of varied selections.

All too often, new restaurants will list numerous menu selections simply to round out the menu or offer token items that are on almost every menu. Your bakery should move toward specializing and serving only those menu items which they can prepare *better than the other establishments* in the area. Creating a diversified menu for the sole sake of offering a multitude of items makes it extremely difficult to make these profitable.

Specialization in the bakery business is the key to building a solid reputation. Word of mouth is the most effective form of advertising available to the bakery manager. A successful menu is one that is honed to build a reputation for excellence.

Limiting the menu will create advantages for the entire operation. The kitchen staff will become more experienced and skilled at preparing each item. Your service staff can then concentrate on promoting and recommending those items which the bakery specializes in. From an administrative standpoint, a smaller menu will be easier to control. Purchasing will center around only a few major food products; thus the buyer may utilize his large purchasing power to obtain price breaks, discounts and above-average service.

Once the menu is finalized, it will be necessary for you and your staff to become thoroughly familiar with every aspect of each menu item. Extensive experimentation in the kitchen will be needed to discover the precise recipe ingredients, amounts and preparation procedures. Take the time to find out everything there is to know about the menu items. Determine where the raw ingredients come from; which is the best type or brand to purchase; and how the kitchen staff can best handle and store these products.

Portion Control = Savings

The rule for developing a portion size is to use the largest portion feasible but charge accordingly. It is far better to serve too much food than too little. The crucial element, which must be constantly reinforced, is that every serving must be a specific weight and size. Portion control is the foundation of a good cost control program. Its importance cannot be overstated.

Portion-controlling all food items is an effective way to control food costs, but it also serves another important function. **It maintains consistency in the final product.** Once the precise recipe is developed, the completed menu item should look and taste exactly the same regardless of who prepared it. Customers won't be unhappy because the sandwich they ordered on Tuesday is much smaller than the one they ate the prior week.

Portions may have a small predetermined size/weight variance. Determine the proper weight, volume or other measurement for each ingredient and or finished item. For example, if the set portion size for the roast beef in a sandwich is 10½ ounces, the sliced meat may range from 10 to 11 ounces. Any amount over 11 ounces must be removed. In fact, many food service establishments allow a variance of only 1/8 of an ounce!

Since portion-controlling is such a vital kitchen function, purchase the best scales and measuring devices available. A good digital ounce scale will cost upwards of $200. However, this investment will be recouped many times over from the food cost savings it will provide. Purchase at least two ounce-graduated scales for the kitchen and always keep a third available in reserve. One floor-type pound scale with at least a 150-pound capacity will be needed. This scale will be used to verify deliveries and raw yields. All scales should have a temperature-compensating device. Maintain these scales per the manufacturers instructions; clean them periodically and oil when necessary, and they will provide years of service. To ensure the accuracy of the scales test them periodically with an item of known weight. Most good scales come with a calibration kit.

If final food assembly is done at the front counter, have your prep staff weigh, measure and package components for specific products. In the above example, packets of sliced roast beef would be wrapped in plastic ready for the front counter sandwich maker.

For practical reasons some food items, such as dressings, sauces and butter, are portioned by weight. However, they should still be portion-controlled by using proper-size spoons and ladles. Soups and condiments must be placed in proper-size serving containers.

At each work area of the kitchen, place a chart listing the portion sizes and other portion control practices. All employees must use the measuring cups and spoons and the recipe manual when following recipes. Remember, that the basis for the food cost program you are developing is based upon the knowledge that every item has a precise portion size. Your responsibility to ensure that these standards are being practiced and adhered to.

Recipe and Procedure Manual

Your **Recipe and Procedure Manual** will contain all the bakery's recipes, preparation procedures, handling instructions and ordering specifications. This manual, if properly used, will ensure perfection and consistency every time the product is prepared.

The Recipe and Procedure Manual must be available to the kitchen personnel at all times. Recipes should never be prepared from memory. The employee, without constant reenforcement from the manual, will tend to forget the exact proportions, and may even eventually leave an entire ingredient out of the recipe. For this reason and to ensure consistency, the Recipe and Procedure Manual should be open and in front of anyone preparing a food product. The pages of this manual should be sealed in plastic to protect them while in the kitchen.

Assembling sandwichese and other freshly prepared food items won't require an extensive recipe; however, a small laminated card with weights and measures should be posted in prep areas.

Ordering Manual

The **Ordering Manual** contains all of the products that will need to be reordered, the minimum and maximum inventory levels desired, vendor and a request to order form. The completed Recipe and Procedure Manual will list all of the food items you will need to order. Simply transfer these food items onto the order forms in alphabetical order. It is a good idea to group similar food items together, such as dry goods, seafood, poultry, dairy products, produce and so forth.

Projecting Menu Costs

In order to accurately assess the price you must charge for a menu item, you must know the exact food cost of that item. Projecting menu costs is simply a matter of mathematics.

You will need the completed **Recipe and Procedure Manual** and the current price lists from your purveyors. From your sales representative, obtain projections on the average yearly prices for the major food items (such as meat and cheese) you order.

Using the Recipe and Procedure Manual and the current price lists and price projections, compute the cost of each recipe item and place the amount in the column under current cost. Prices must be broken down into fractions; round all the amounts off to the nearest cent. Should estimates need to be given, it is better to figure a little high in order to cover yourself.

Don't forget to factor in the condiments and accompaniments not listed approximately 5 percent of the total entree cost should cover these expenses. Adjust accordingly for each entree. Once open and operating you will be able to fix an average cost per customer for all miscellaneous costs including free beverage refills.

When computing the portion costs, you incorporate the cost of normal waste created during cleaning, cutting and preparing meat and produce. The amount of usable portions you get when you're finished trimming or cutting a piece of food is the yield.

To Compute The Yield Percentage:

1. Compute the gross starting weight in ounces.
2. Compute the net ending weight in ounces. (This is the number of ounces yielded after the item is portioned.)
3. Divide the net yield (in ounces) by the gross starting weight (in ounces).
4. The resulting figure is the *yield percentage*.

To Compute The Actual Portion Of A Product:

1. Divide the price per pound by the average yield percentage. (This is the actual price per pound after waste)
2. Divide the actual price per pound by 16 to get the actual price per ounce.
3. Multiply the actual price per ounce by the average portion size: this figure is the actual portion cost.
4. Total all the current costs for each item. This figure is the estimated total portion cost.

This cost figure is, of course, not completely accurate because of the large number of variable factors used in the computations. However, this figure is an educated estimate from which you may accurately set your menu prices. Remember that the costs listed here are food costs only; no other costs (such as labor, paper products, plates, etc.) are factored in at this point.

Projecting the Actual Average Cost per Customer

Once set up and operating, it will be relatively easy to compute the actual average cost per customer. The actual average cost per customer should be projected once every month. This ensures that the estimates used in computing the menu costs are accurate. Also, this is an extremely important procedure for bakerys that offer "help yourself" items such as beverages and salad bars. Bakerys offering "all you can eat" specials must project their actual average cost per customer at least once a month, or better yet, biweekly.

To Project The Actual Average Cost Per Customer:

1. Keep a list of all the food items you do not charge for during a specific test period and their prices. You can develop this list from the invoices which detail daily purchases.
2. Add into this figure the dollar amount of food you have on hand at the beginning of the test period. This pertains only to the food that you are not directly charging for.
3. At the end of the test period, subtract the amount on hand from the total.
4. Divide the total cost by the number of customers served during that period. This figure is the average actual cost per customer. Use it in projecting menu costs instead of other estimates.

Projecting Menu Prices

Projecting menu prices is a complex procedure because of the number of factors must be considered. In order to have a profitable "prepared meal profit center", you will need to keep your food costs at or in the 35–45 percent of sales.

The food cost percentage is the total food cost divided by the total food sales for a given period. For example, if the total food sales for a given period was $100,000 and the total food cost was $40,000 for that same period, the kitchen would be operating at a 40-percent food cost of sales. One percentage point in this example would be worth $1,000.

Computing what you must charge for each entree item is relatively easy. You will need the estimated total portion costs from the preceding section. The total portion cost (food cost) divided by the menu price (food sales) must equal a food cost percentage of between .35 (35 percent) and .40 (40 percent).

Portion Costs ÷ Menu Price x 100 = 35–45%

Simply plug different menu prices into the formula until you reach the desired food cost percentage.

Be aware that using this formula on per item basis can give you some undesired results. Some of the prices you'd need to charge would be simply too high. No one would ever purchase the item at that price. What you must do in these cases is balance out the menu with high and low food cost items. The average cost of the menu must then be in the food-cost-percentage range desired. Some items will usually have a lower food cost percentage (and will be your most profitable). Promote these lower food cost items to offset the higher ones.

Find out what other small delis, cafes and coffee shops in the area are charging for similar menu items. Your customers will dictate what the market will bear. The bakery manager must set his menu prices based on what his customers will spend and what he must charge in order to make the desired profit margin.

Appetizers, side orders and beverages can be priced at a very low food cost percentage. These items will contribute to a large percentage of your food sales and will lower your overall food cost percentage. Some bakery managers, realizing this important point, have set up promotional contests awarding prizes or money to the waiter or waitress who sells the largest percentage of side orders. This can be very effective if your staff do not "hard sell" the items, but rather suggest the accompaniments to their customers.

Maintaining the food cost percentage is critical to maintaining your profit level. However, the food cost percentage does not tell the entire story. You must also be interested in getting the largest guest checks possible, to bring the largest percentage of gross profit to your bottom line. For example, which would you rather sell:

a) an item that sells for $5 and has a food cost of 35 percent
  (a gross profit of $3.25), or
b) an item that sells for $10 and has a food cost of 50 percent
  (a gross profit of $5.00)?

Here is an example of a higher-food-cost-percentage item actually bringing a higher gross profit. Consider this important point when pricing out the menu!

If you are not reaching your food cost goals or are not getting as high a check average as you'd like, it may be because of your menu descriptions or design. Not all items on your menu can be low cost and high profit. Your menu is likely a mix. Your menu design may be emphasizing high food cost or low profit items. Fixing this will help decrease food cost and increase profits. Remember, if you sell too many high cost items your food cost will go up, because many of these (such as beef and seafood) have a high cost as well. On the other hand, if you sell too many low cost items, your check averages and gross profits will decline. Keep in mind in creating your menu, you want to have a sales mix of both these types of items.

Preparing Your Menu

Your menu may be printed and laminated, be a simple flyer, be posted on the Web, found only on a in-store sign or be included in a Delivery Service guide. No matter where your menu and bakery product listings are displayed some basic guidelines should be employed. In every medium except the Web, space will be limited.

Truth and Accuracy in the Menu

Careful consideration must be taken when writing the final menu to ensure its complete accuracy. Few bakery managers would purposely deceive their customers, as the bakery would only suffer in the long run. However, you must become aware of the unintentional inaccuracies you may have in the menu and the governmental regulations regarding this.

Providing your customers with a tantalizing menu is important; however, you must accurately represent your food items.

- Is your 3-egg omelet really made with three eggs?
- Do I know that my organic-produce is truly certified organic?
- Am I clear on how coupons or promotions can be used?
- Are there extra charges for "sharing plates" or making substitutions?
- Do I tell people I serve generic soda when they ask for a Coke®?
- Do I call it Roquefort® cheese when it is really blue cheese?
- Do I clearly identify ingredients that are common allergens?
- Am I clear as to points of origin such as Florida orange juice or Oregon bay shrimp?
- Do I use the term "fresh juice" only for a juice without additives and prepared from the original fruit within 12 hours of sales?
- Are food items/entrees that deviate from common preparation, ingredient or service style noted? Does your New England Clam Chowder follow the classic recipe or have you made changes?

- Instead of using the term "homemade," do I use more accurate terminology, like "homestyle," "homemade style," "made on the premises," or "our own"?
- If I use any of the following terms, am I sure I can substantiate them?
  o   Fresh daily
  o   Fresh-roasted
  o   Flown In Daily
  o   Kosher Meat
  o   Black Angus Beef
  o   Own Special Sauce
  o   Low Calorie
- Am I careful not to misrepresent fresh orange juice as frozen?
- Do I accurately use the commonly accepted means of preserving, such as: canned, chilled, bottled, frozen and dehydrated?
- Am I always absolutely accurate in the terminology used to describe the method by which the food is prepared? Is it really charcoal Broiled, fried in butter or mesquite-smoked?
- Do my menus, wall placard or other advertising materials containing pictorial representa-tions always portray the actual product with true accuracy?
- Am I sure I never risk the public's health by misrepresenting the dietary or content of a food?
- Do "salt-free" or "sugar-free" mean just that?

Disclaimers

Many food service establishments, in order to limit liability, will print what is called a disclaimer. A disclaimer is simply a statement that what you have printed is accurate to the best of your knowledge but that the bakery cannot be held responsible for any actions beyond its immediate control. Here's one type of disclaimer:

We serve only the finest food available. However, at certain times during the year we may not be able to obtain the exact product desired, therefore we may substitute a similar product that will be equal or superior to the original item. Should this be the case our staff will inform you of the substitution.

Carefully analyze your menu for possible misrepresentations. Self-regulation in the food industry is the key to maintaining the high standards and reputation it enjoys.

Nutritional Claims

If you want to include menu items that are marketed as healthy (i.e., heart-healthy, low-fat, reduced-fat, cholesterol-free, etc.), make sure you have the nutritional information for these items readily accessible. Items described as "fresh" are included in this category.

Basically the FDA regulations state that if you make health/nutritional claims on your menu you must be able to demonstrate there is a reasonable basis for making them. There is some flexibility in how bakeries may support the claim, but basically they must be able to show customers and officials that their claims are consistent with the claims established under the Nutrition Labeling and Education Act. Additional resources on labeling and organic claims are in Chapter 24.

Some establishments are beginning to list ingredients and "Nutritional Facts" labels on the menu for the convenience of their customers. Such a label indicates the item's value in calories, total fat, cholesterol, sodium, carbohydrates, protein, etc. Some states now require food purchased for take-out to be labeled in this manner.

If you decide to include items on your menu that will have health claims, you must decide on the best way to communicate the nutritional information to your customer. Here are some things to consider:

Information should be correct and clear. You may not need to include all the nutritional information on the menu; you may only need to have it available. For example, many of the fast-food chains simply list this information on a poster in a public area in the bakery. Depending on your menu format, you may not want to clutter the space with this information. If you find that your typical customer is requesting this information, it may be better to include it on the menu.

Displaying and Printing Menus

Attractive and informative menus are not just lists of facts – they are valuable sales tools. The menus in bakerys across the country are probably more diverse than the food itself. Menus range from freehand writing on a white piece of "8 ½-by-11" paper to menus printed on boards, tables, walls and Web sites.

Your menu can be easily turned into a promotional vehicle for your bakery. Your menu should communicate to your customer your objectives (freshness, handmade) and identity (brand name). Your menu design will directly affect customer choices and check averages and your profit goals. A well-designed menu can attract a customer's attention to specific items and increase the chances that the customer will purchase those items. For instance, if you put an item in a box on the menu, the customer's eye will be drawn to this area of the menu.

Regardless of how creatively the menu is utilized, it should be typeset and either printed by a professional or with the professional menu software (discussed below). Simply using an unusual type style will dress up any menu. Just make certain that it is easily readable and large enough type. Discuss the possibilities with your local printer or graphic-art person.

Menu Design Help

You can also find companies specializing in menu production for more formal menus. In a bakery, people don't expect your menu to look like it belongs in a dinner house so don't overdo it!

Artwork should be used if at all possible; use the bakery's logo if nothing else. Your local printer may have an artist on staff or know of some freelancers in the area who can help.

- Microsoft Office includes generic templates that can be used. Additional templates and graphic art (including menu-specific options) are available free at **www. office.microsoft.com**.

- Restaurant MenuPro software is available from Food Software.com (**www.foodsoftware.com**).

- Soft Café offers menu design software at **www.softcafe.com**.

- Custom menu covers can set you apart. Check out Impact Enterprises at **www.impactmenus.com**.

Chapter 12

Recipe Development and Production

eveloping, testing and producing your baked goods and menu items is a process of trial and error. For baked goods, you'll have variables such as altitutude, humidity, temperature and every ingredient that affect the final product.

Designate your entire baking staff as your research and development department. Let them explore all the variables, document their findings and create a taste test.

Experimentation

Experimenting with all the variables will help you isolate the best components for a quality product at a fair price. You may not be seeking perfection but a good compromise so your customers, your employees and your profits will benefit. A finished product may look beautiful and taste fantastic – but if you have to sell it at twice the going rate – you probably need to rethink it. Likewise, it may be wildly popular with customers but you have to hire a prep person – leaving you with a no-profit item.

- **The right ingredients** – Which flour do you use? Walnuts or pecans? Meringue or whip cream?

- **The perfect formula** – Basic ingredients combined in exact proportions.

- **The optimum cooking temperature(s) and time** – With the advent of computerized ovens, the possibilities are virtually endless. Do you start at 450° for 5 minutes, then 375° for 20 minutes? How about 10 minutes with the fan on and then 35 minutes with no fan?

- **The best pan** – Should you use a wide, shallow pan or a tall pan?

- **The fastest prep** – Should you purchase prepared lettuce or chop it up in the kitchen?

- **The beautiful appearance** – Brown enough? Tall enough?

- **The taste** – Melt in your mouth good. It tastes as good with the less expensive brand.

Ask for Help

An experienced staff can be invaluable when you need help in tweaking your recipes. You'll also find useful guidance from peers around the world through bakery associations and food service industry websites. In addition, your vendors and suppliers can be excellent resources for product testing.

- **Equipment manufacturers** – Visit their test kitchens or have a sales rep. spend time with your staff during development. The computerized oven manufacturers are especially helpful in programming their equipment for an optimum bake.

- **Utility Company** – Many gas and electric utility companies have complete test kitchens filled with a variety of cooking/baking equipment. Considering a convection oven? Bring your dough samples in and try one out.

- **Food suppliers** – Have them bring you samples (enough to make one or two small batches) of various flour grinds/blends to test. Their representatives often have baking experience and other resources that they can share.

- **Water Company** – Have them test your water and make recommendations on how to deal with particulates, hardness and pH factors which may be causing havoc with your recipe.

Test Before You Commit

Introducing new products or menu items may mean you'll have to invest in new ingredients, special equipment and then you'll have to update signs and reprint menus. Before rolling out a new product or menu item:

- Offer it as a daily special a few times and have customers complete a survey card.

- Iinclude a self-addressed prepaid postcard with questions when you pack up take-out orders.

- Pass out samples and ask for input. What do you think? Would you buy it? Would you order it for lunch? How many would you buy at a time?

What Will Be On Your Shelves?

As mentioned early, your chosen recipes may be a compromise so that your customers receive the quality they want, your employees can quickly and easily product sufficient quantity and you can charge a fair price. As with most businesses, a small percent of your products will account for the majority of your revenues. Another group will be in the mid-profit range. If your margins are strong, you may even consider including a few low-profit items if you feel that they will do one of two things; 1) Bring new customers in; 2) Keep customers happy.

Setting Production Standards

Once you have determined the best products for you to produce, you'll need to create a production standards. These standards are levels that you establish for:

- **Volume by hour/shift/day** – benchmarks for productivity.

- **Quality standards** – define what is optimum, acceptable and "toss it" for each batch.

- **Acceptable substitutions** – define what if any ingredient, pan or cooking substitutions are allowed.

Chapter 13

Equipping Your Bakery

Your equipment and tools are an investment in your ability to produce quality products quickly and efficiently. Equipment has a "return on investment" factor (ROI) that makes it *pay for itself* over time; the shorter the repayment time, the better investment it s for your business. The ROI can be calculated in dollars (increased sales, fewer employees) and time (faster production). Other important benefits are often difficult to put into comparable numbers such as improved quality (happier customers), reduced risk (reduced injuries and stress) and competitive advantage (exclusivity, enhanced service, more choices).

Equipment Budgeting

The No. 1 question is *"How much should I spend?"* Quality and pricing levels vary widely so there is easy answer. We do know that successful bakeries won't spend more than they need to.

For some light-duty equipment, a less expensive, yet highly serviceable brand may be the best choice. While, heavy-use equipment may require the best quality available for lasting performance.

How do I keep within my budget?

- Develop an equipment/fixture/tool wish list.

- Divide your list into three priority categories.
  - o Cannot Live Without
  - o Would Make Life Easier
  - o Wouldn't It Be Great

- Allocate your budget primarily to the first category. This is the equipment that makes you money.

- Review the items in category two for potential time and money-savings.

- Be very objective about items in category three.

- Will the $14,000 Espresso machine make a difference in your bottom line?

- Analyze your second and third category items for their potential ROI.
  - o How long it will take to pay for itself?
  - o Will it make your money or just make you look better?
  - o Is leasing a wise alternative?

- Review every decision from your CPA's viewpoint. Buying cooking equipment can be like getting a new toy! Don't let your excitement or a salesperson's pitch eat up your budget.

- Work closely with a food service consultant or do your own research. Compare features and benefits to your acquisition and maintenance costs.

- Negotiate for a better price. Start by asking 50 percent off the list price. Depending upon the equipment and the dealer's purchasing power, there is almost always some negotiation room.

- Shop with major bakery/restaurant supply houses first. Factory discounts to volume distributors could give you some additional negotiating room that isn't available with smaller suppliers.

- Ask about last years' models. Incentives may be available on older or overstocked models.

- Check the Web for discounts. Search under "Bakery Equipment" and "Restaurant Equipment." Often equipment is dropped shipped from the factory, so freight may not be a factor. Ask about sales tax, as this is a "gray" area for online purchases. When in doubt, Google it.

Buying Used Equipment

Buying used bakery/restaurant equipment and fixtures can be a very wise decision. Just like a used car, equipment depreciates more during the first year or two. With the food service industry's high failure rate, there's always plenty of "almost-new" equipment available.

- **Before searching for a used piece of equipment, shop for new.** This will give you a benchmark of features, quality levels among manufacturers, and pricing. Just as a used Mercedes is a safer investment than a used Yugo, focus on the top manufacturers with a reputation for quality.

- **Ask about the repair history of the make/model.** Institutional equipment typically has a long projected lifespan. Your dealer will probably have personal experience with the equipment.

- **Learn the terms "reconditioned" and "rebuilt."**
  o   Reconditioned equipment – cleaned, worn/broken parts replaced, short dealer warranty. Typically priced at 40 to 50 percent of new.
  o   Rebuilt equipment – totally dismantled and rebuilt, longer dealer warranty. Should perform equal to the manufacturer's specs. Typically priced 50 to 70 percent of new.

- **Verify the equipment's age and history.** Use manufacturer serial numbers and service records to check age and care. Don't rely on an "only driven on Sunday's by Grandma" story.

- **Ask the used equipment supplier about their trade-in policy.** Some suppliers will give you above average trade-in values when returning to purchase a new version.

- **Shop for used equipment online (auction and direct purchase sites), at bankruptcy auctions, from new equipment dealers, and food equipment Rep. Groups.**

- **Ask if they have demo models available.** Trade show, showroom and test kitchen models typically can few "miles" on them and reduced "scratch 'n dent" prices.

- **Save *time* by buying used.** Lead times on new equipment can be lengthy.

- **Don't buy...**
  - o Cosmetically damaged equipment/fixtures that will be visible to customers.
  - o Anything with moderate (or worse) rust (except restorable cast iron).
  - o "Married" equipment (where the legs from one model have been attached erroneously to another model).
  - o Foreign-made equipment that wasn't made specifically for the U.S. market. Unknown electrical conversions can be a problem.

Leasing Your Equipment

For some bakery owners, leasing can be a way to extend your available capital. Leasing is 100% financing. Depending on your lease, you may receive better tax benefits at lower monthly payments while preserving your working capital and preserve borrowing capabilities.

To help you determine what financing method suits your needs, here are some helpful ideas and resources.

- **Don't think of leasing as easy money.** The *true* cost of leased equipment can be much greater than the purchase price. You are paying interest even when you lease.

- **Avoid personal guarantees if possible.** If you sign it, you're liable even if your bakery closes or the equipment doesn't last.

- **Educate yourself about leasing before shopping for a leasing company.** Leasing companies pull credit reports. Too many inquiries can negatively affect your credit reputation and lower your rating by making it appear that you are always looking for new loans..

- **Confirm who is responsible for service and maintenance.** The manufacturer's warranty extended to the lessee. However, in most cases, you are 100% responsible for keeping the equipment in good working order and resalable condition.

- **Compare your total annual lease costs to your annual depreciation benefits.** Food service equipment is often reported have around a 7-year depreciation rate as compared to typical leases are 36 to 60 months. It is still important to note though that depreciation rates can vary from year-to-year, so it is always important to be aware of the constantly changing market in which you are entering.

- **Don't lease** items with a short life or items that are fully deductible the purchase year such as flatware, glassware or dinnerware.

- **Be aware of leases with low buy-out provisions.** The IRS may classify it as a purchase agreement subject to depreciation rules instead of a 100% expense.

- **Make certain your insurance adequately covers leased** equipment for fire, theft or other losses.

- **Get the fair market value information in writing.** Equipment with unrealistic residual values can have excessive buyouts. Check the used market for comparison figures.

- **Read the lease before signing.** A lease is a legal contract! You might even have your lawyer review the fine print.

- **Estimate your monthly payments** and learn more about leasing from ELFA (Equipment Leasing and Finance Association) at **www.elfaonline.org**

- **Ask about used equipment leasing.** This can be a cost-effective way to obtain the most expensive equipment.

Making Wise Equipment Purchases

Your job in choosing equipment is to get the best value for every dollar spent. We've outlined some helpful suggestions to ensure that get the quality, service and performance you pay for and need.

- **Seek out recommendations.** Unfortunately, there is no *Consumer Reports* for restaurateurs. But asking peers, used equipment dealers, industry association members and foodservice equipment specialists can help you learn about desired features, life expectancies and brand names to consider and/or avoid.

- **Contact your local gas and/or electric utility company.** Many utility companies have fully outfitted test kitchens where they promote gas or electric equipment from major manufacturers. Ask about available rebates and promotional programs.

- **Don't overlook custom-built equipment.** To get the quality, service and performance you need, the solution may be custom-built. This can be the best choice when looks are important, you have unique specifications or your usage exceeds the capacity of stock equipment.

- **Comparison shop.** Have your consultant or equipment dealer give you good, better, best recommendations. Compare features, operation costs, and life expectancies.

- **Establish substitution rules.** Sometimes the equipment you select is not available due to excessive lead times, product discontinuation or unforeseen price increases. Carefully examine substitutions for suitability.

- **Look for commercial-grade materials and superior construction.**
  o Choose high-grade stainless with welded joints.
  o Verify that the gauge of steel (remember the smaller the number the thicker the steel) used is as quoted.

- **Don't have equipment delivered until you are ready to install it.** You risk dents and dings. Dust can irreparably damage fragile equipment.

- **Choose equipment doors that open away from the nearest worktable to facilitate removal of hot and heavy pans.**

Which Quality Level?

Should I invest in the top brand on the market or purchase a serviceable low-end model? Who wouldn't want the latest and greatest of food service, office and business equipment? Investing wisely is a contributing factor in your long-term success.

We suggest that you carefully weigh your emotional, creative and business needs in assessing whether a specific piece of equipment, tool or fixture serves your immediate and future needs.

- **Ask the following decision-making questions.**
  o Does it meet my budget?
  o Does smaller, more efficient equipment save precious space?

o   Will my food or service quality improve with this equipment?

o   What is the ROI?

o   For specialty equipment, will I sell enough to pay for it?

o   Will it save energy costs, reduce overhead, or make employees more productive?

o   Will it be on display where looks are important?

o   What type of routine maintenance does it require?

o   Is local service available and affordable?

o   Is it difficult or expensive to operate daily?

o   Is an economical service and/or maintenance contract available?

o   Is it the most productive and energy efficient equipment for the job?

o   Is the lifespan greater than the payment or lease terms?

o   Are there trade-in/trade-up programs available?

o   What's the resale value if I need to sell or trade-up?

o   Does it meet sanitation, plumbing or building code requirements?

Service Contracts

Many gas/electric companies and equipment manufacturers have service contracts that may be purchased. If available, it is highly recommended that you purchase them. Equipment that is maintained to the manufacturer's specifications will last longer and operate more effectively and efficiently. One particular item that should be in any service contract is the calibrating of the baking ovens. It is critical that all ovens register their temperatures accurately for consistent cooking results. Heavy-use ovens need to be calibrated every month.

Equipment Records

Set up a loose-leaf binder to contain all the information on your equipment and its maintenance schedules. Included in this binder should be warranties, brochures, equipment schematics, operating instructions, maintenance schedules, part lists, order forms, past service records, manufacturers' phone numbers, a chart showing which circuit breaker operates each piece of equipment, etc. Keep this manual up to date from the very beginning. Become aware of your equipment's needs and act accordingly. Train your employees thoroughly in the proper use of your equipment and it will serve the bakery well for many years.

Bakery-Specific Equipment, Tools and Supplies

The most important equipment for a bakery is a mixer, mixing benches, proofer and high capacity ovens. In addition, you'll need equipment, tools and supplies for slicing, decorating and packaging. The type

and capacity of your equipment should be based upon your anticipated hourly production volume. Operating some equipment for long periods at its maximum output can decrease its life and efficiency and increase repair and maintenance costs. Invest in equipment where you are working in the middle of its capacity range. This means you won't be "overworking" the machine and will be able to increase your output without purchasing new equipment.

Whip It Up

There are a wide variety of prep equipment available for bakers. By automating some functions, you can improve your portion control, create a more consistent product and reduce production labor. You can find equipment that will handle virtually every aspect of putting a recipe together.

Scales

Most commercial bread recipes rely on weight in measuring raw ingredients. Weight is more accurate and easier to calculate when increasing or decreasing the recipe size. Weighing dough also ensures consistency in your bake times and output. Different types of scales can used throughout the bakery.

- At the delivery door, a floor scales to weigh incoming bulk items.

- Baker's scales and dough scales that weigh ounces and pounds should be next to all prep areas.

- In your decorating/finishing areas, add a scale to weigh frosting ingredients for limited quantity icings and to weigh out buttercream frosting or filling before tinting or flavoring.

- Some smaller items, such as cookies, can be sold by the pound, so a front counter scale will be needed.

- In sandwich or bagel prep areas, a small scale will help control the costly ingredients (meat, cheese, etc.).

Digital scales can be a wise choice as they are faster to read, more accurate, and many display the weight even after removing the bag or bowl. All scales require periodic calibration to ensure accuracy. When investigating a supplier, ask about maintenance and calibration service.

Cut Them, Drop Them, Fill Them

Cookies, muffins, donuts, scones, and other "piece" or filled items can all benefit from tools that measure dough, fillings and frostings. These are a must for saving time and producing a consistent product. Review manual tools, semi-automatic tools and fully automatic equipment. Your volume requirements will be the deciding factor for these.

Specialty Equipment

Donuts, bagels, croissants and other specialty breads typically require specialized production equipment. From dough handling to shaping to frying, you'll find a specialized piece of equipment to streamline the task.

Specific task ovens are those that you would use to produce specialty breads such as:

- Artisan breads baked in a wood burning oven from Earthstone Ovens (**www.earthstoneovens.com/commercial.html**)

- Pita bread baked in a high-temp oven from Bramco (**www.bramco.co.nz**)

- Crusty French loaves baked in an oven outfitted with steam injection. Several major manufacturers have this feature in combination with convection or standard electric cooking

Donut equipment:

- General-purpose commercial fryers can be used for frying donuts. Fryers marketed as "donut fryers" are also available from major manufacturers such as Vulcan Hart (**www.vulcanequipment.com**), Lang Manufacturing (**www.langworld.com**) and Blodgett (**www.blodgett.com**).

- Belshaw at **www.belshaw-adamatic.com/** (everything from cutters to finishing tables).

- Baxter at **www.baxtermfg.com** (glazers).

- Houpt manual cutters available through your local restaurant equipment supply house.

Bagel-forming equipment:

- Thompson Bagel Equipment at **www.bagelproducts.com**.

- ProBake at **www.probake.com**.

- Oliver Products at **www.oliverproducts.com**.

Croissant equipment:

- Dunbar Systems at **http://www.dunbarsystems.com**

Mix It Up

Commercial mixers are a necessity for mixing dough, custards, whipped cream, frostings, etc. A counter mixer located in your decorating area can be useful for small batches and for invigorating refrigerated decorative frostings.

The industry leader in production mixers is Hobart. (**www.hobartcorp.com**). Their website has calculators to determine capacity needed, potential cost savings when using auto-scrapers and advice. Their planetary action mixers are available in counter and floor models.

Your local restaurant supply house can assist you with determining the right mixer for your needs. Your mixer will be a heavy-use item so invest in the best.

Heat It Up

Proofers are essentially sealed boxes that keep yeast dough at an optimum 98 to 100 degrees in a moist environment. Proofers can be "closets" where entire racks can be wheeled in and out or they are stand-alone units where pans side onto racks. Manufacturers such as Baxter offer retarder (to slow rising process)/proofer combinations offer cold and hot temperatures for holding product for future baking.

Proofer/oven combinations feature a deck oven with heat/moisture capabilities for proofing. NU-VU (**www.nu-vu.com**), Doyon Baking Equipment (**www.doyon. qc.ca**), Empire Bakery Equipment (**www.empirebake.com**) and others manufacture these space-savers.

Baked Fresh Daily

Ovens come in hundreds of configurations powered by gas, electric or wood. Units can be regulated by high-tech computerized systems or by hand. Your space limitations, product requirements and budget will all be deciding factors on purchasing an oven. Beyond basic construction quality, the two most important oven considerations are capacity (rack sizes, adjustability, pan sizes) and temperature control (accuracy and stability).

Multi-purpose ovens come in a variety of configurations. Many manufacturers have test kitchens where you can see actual product being baked and even take dough samples to test. Contact your local utility company, restaurant equipment supply house, food service equipment representative group and favorite manufacturers to locate a facility near you.

- **Revolving tray ovens** – Sheet pans and trays move in a revolving motion to promote even baking

- **Convection ovens** – Hot air is circulated through the oven to speed baking and keep temperatures level throughout the entire oven. Can be deck or rack configuration.

- **Deck ovens** – Flat steel or stone surface oven. Single or multiple decks in varying heights are available.

- **Rack ovens** – Multiple sliding rack configurations for volume baking with same temp/time requirements. Offers the best use of floor space.

For high volume operations, there are tunnel, conveyor and traveling tray ovens that move product through the oven. These ovens require significant floor space and require professional installation.

Fix It Up

The right decorating and finishing equipment and supplies can transform an ordinary sheet cake into the centerpiece of a celebration party. Basic cake decorating items are not expensive – just make certain that have plenty of each so that you don't waste time. Equipment such as airbrushing equipment and projectors to reproduce photos on cakes are a way to tap into your creativity and produce a unique product. Research the latest products and time-savers at:

- Pastry Chef Central (**www.pastrychef.com**)

- Candyland Crafts (**www.candylandcrafts.com**)
- Kopykake (**www.kopykake.com**)
- Lucks Food Decorating Company (**www.lucks.com**)

Package It Up

You'll need a variety of packaging products, depending upon how your customers purchase and carry specific goods. Some items will do best if packaged in plastic bags; however, you may find that paper products are often better.

Plastic, Please

Sliced bread, rolls and bagels are suitable for food-safe plastic bags. Very few artisan breads can handle the artificial environment created by plastic as their crusts need to breathe. Create custom-imprinted bags that feature your bakery logo, some sales-oriented product information and the required ingredient disclosure information.

Paper, Please

Bread bags enhance the beauty of handcrafted artisan bread. Whether you want an unbleached rustic look or a classic white bag, you should match your bag to your bakery décor/theme/style. A variety of bags and protective packaging should be purchased. Ask your supplier to visit with samples of standard sized bags (using standard sizes is more cost-effective; however custom bags are also an option). Have product samples available to select the fewest bag configurations that will fit the most products.

Cake boxes need to be tall enough to accommodate your tallest cake and sturdy enough to be carried home without damaging the dessert. Select one or two flat boxes to accommodate sheet cakes, tarts, tortes, pies, donuts and other items that should not be stacked.

Other Serving and Carry-out Packaging

Review your menu to see what other items people will be carrying away from your bakery. Coffee cups and hot sleeves, candy boxes and quantity orders all need specialized paper packaging. Potential vendors can help you make the best selection for you and your customers.

Product Labeling

When choosing packaging you must also consider your labeling requirements. You'll find a brief discussion on nutritional, organic and other food labeling in Chapter 13.

Your best sources for current product labeling requirements (state and federal) are the various food service, retailer and bakery associations.

Wash Up Afterwards

Proper sanitation depends on a washing system that protects your customers while being efficient and cost-effective. Whether your establishment only requires a multi-sink configuration or you need the performance of a conveyor system, your dishwashing decisions directly affect your bottom line.

Sanitation occurs when water, chemicals and heat are properly combined. Your basic needs in dishwashing are: waste removal, washing, sanitizing, rinsing and drying. Your first step is to familiarize yourself with your local health department codes.

- **Make certain water pressure is ample to operate your chosen commercial dishwasher.** Poor water pressure may slow cycle times, inhibit automatic settings and not meet sanitation standards.

- **Review your hot water capacity and recovery times.** You'll need temps of 150º and above depending up your usage and/or machines capabilities.

- **Invest in a hot-water booster.** Check with your local utility company regarding subsidies and rebates on select equipment.

- **Calculate your water hardness.** Rinse aids may not be sufficiently effective without a separate water softener.

- **Determine what you need to wash and when.** Will quick turnaround of expensive stemware require a separate glass washer? Will you require a large storage area for soiled utensils or pots that are only handled after the rush is over? Will you have unique shaped equipment that will require soaking?

- **Install a low-flow (1.6 gpm) pre-rinse nozzle** at your dishwashing station and save up to $100 a month in energy, water and sewer costs.

- **Research equipment rental/chemical purchase programs.** The supplier owns and maintains the equipment. You pay a rental fee and purchase the supplier's chemicals. Auto-Chlor System, a well-known national supplier, can be found on the web.

- **Create ample, convenient (with no potential cross-contamination) and secure storage for chemical dishwashing agents.** Rental/supply programs with monthly service require less chemical inventory.

Non-Automated Washing

A multi-sink configuration that complies with local regulations is probably you best option to clean oversized pots and equipment components. Be certain to…

- Add ample counters and drain racks. Local codes dictate widths and lengths.
- Separate dirty and clean dishware to avoid contamination.
- Install a detergent dispenser to reduce waste.
- Maintain a consistent and high enough water temperature by placing sinks near a dedicated water heater.

Refrigerators & Freezers

Your chilling requirements may mean you need more refrigeration than a typical food service establishment. Walk-in units allow you to wheel in product to hold, retard rising and lower temps before freezing. You'll also need refrigerator space to store sourdough starter, yeast, cream, butter and eggs.

When one thinks of fresh baked goods, freezers typically don't come to mind. However, many produce ingredients can be purchased flash frozen for superior freshness even during off-season. The decision on whether freezing is a good or bad thing for your bakery is solely a personal one. If you never use a freezer – market it. If you do – be careful what you promise.

Computers – How to Use Them and Profit from Them

Computers have obviously become valuable tools for the small businessperson. They offer capabilities that were once only available to major corporations. Even the smallest bakery operation can benefit from owning at least one personal computer. This can be a desktop or laptop model capable of running a word processor, spreadsheet, database, the accounting package of your choice and other off-the-shelf software packages that make your business more organized or efficient.

Computers are integrated into every facet of the food-service industry. Computers can:

- Track your sales and purchases.
- Monitor your inventory.
- Increase your purchasing power.
- Maintain accounting records and payroll.
- Develop menus and minimize food waste.

You'll also find information on marketing your business on the Internet in Chapter 8.

A computer can spit out professional-looking quotes, help you build a portfolio of your wedding cakes, put your daily accounting in order and print out payroll checks. Even if you outsource many of these tasks, having a computer to prepare raw data and present it to your consultant is important.

Industry and task-specific computer systems may not look like a computer; there may not be a monitor and the keyboard may look like a cash register. These specialized system focus on specific functions such as ringing up sales or calculating weights and measures for specific recipes. Computerized systems, such as with alarm and fire suppression systems, can actually be lifesavers.

Computers are valuable tools for the small businessperson; however, they can also be a source for headaches. It's thorough training that turns all those chips and circuits into a valuable support system. Investing in technology is a poor choice if you don't likewise invest in learning how to use it. Your local community college, computer store and the Internet in general can provide you with fast-track classes with hands-on training for everything you need.

When purchasing custom software or hardware systems, be certain that the necessary training and troubleshooting options are available, manuals are easy-to-read, and reliable, on-going tech support is provided. Industry- and task-specific computerization is a long-term investment. New employees and new business needs requires long-term support by your vendor so research these companies and programs before purchasing.

Business Computers

Your personal computer can be certainly be used for accounting, inventory control, personnel support, advertising and business correspondence, but it is probably much more advantageous to keep all of your work data on a PC or Mac that is specifically devoted to your business. Laptops may be a convenient option, yet your confidential data is still at a greater risk of being stolen.

Many major manufacturers and retailers offer small-business leasing. An extended warranty and/or service agreement (preferably an on-site agreement) can be an excellent investment as your "business" will depend upon your computer working full-time.

Utilizing The Internet

The Internet gives small businesspeople access to a wealth of information and resources. You can research equipment, place your flour order, compare prices for paper goods and advertise your bakery.

Please also be aware that such wireless connections to your business mean you need to have extra computer security. Both Windows and Mac operating systems have a built-in software firewall (to stop electronic intruders from accessing your computer and data) but an additional firewall service (software and/or hardware) may be necessary to protect your privacy and confidential files.

The large majority of companies now offer online ordering, order tracking and technical support. The ability to automate repetitive actions can be a real time saver. If your typical order stored online, then it takes only a minute to schedule next week's delivery. If you are confident that your target consumer audience would prefer the old-fashioned options of telephone or in-person orders, keep in mind how high Internet services are on your list of priorities.

Banks offer web access to your accounts including transferring funds from one account to another and paying loans and credit cards. Web-based bill paying services through your bank make it easy for you to pay your bills. Individual credit card companies and utility companies offer online bill pay. Check with your creditor for automatic bill paying methods to save you time and keep your account current.

Just be aware of how long it will take for your creditor to actually receive and post the funds. Just because you are paying online doesn't always mean that your bill can be paid on the day it is due. Read the fine print carefully! Other bonuses are that you may be able to use a credit card, you get an immediate receipt to print, checks are never lost in the mail and you contact customer service at your convenience.

Industry Specific Computer Systems

Point-of-Sale Systems

The most widely used technology in the food-service industry is the touch-screen POS (point-of-sale) system. The POS system, an offshoot of the electronic cash register, was introduced to the food-service industry in the mid-1980s and has penetrated 85% of foodservice establishments nationwide. Freestanding and integrated (networked with your accounting system) systems are available in a variety of price ranges; although all are more expensive than a cash register. The touch screen is effortless, requires minimal training and provides extensive sales data.

Understanding the numbers collected by a POS system gives you more control over your inventory, provides time-specific data (what sells during what hours or days) and provides detailed sales reports.

This information will help you schedule employees to maximize customer service and minimize overtime; monitor shrinkage (waste, theft); and identify productive and non-productive profit centers.

Understanding your POS system ultimately clarifies the bottom line, knocking guesswork out of the equation while paying for itself.

Your bakery POS system will be connected to other locations within the bakery depending upon your food offerings. If your bakery offers take-out or dine-in food prepared in your kitchen, a terminal in the kitchen will receive the customer order immediately and end-of-the-shift or end-of-the-day data sent your office computer. If you are an over-the-counter retail bakery, only your accounting system will need the data.

Some benefits of using a POS system:

- Increases sales and accounting information.
- Offers custom tracking and reporting.
- Reports product sales breakdown for forecasting.
- Shows peak and slow periods for better staffing projections.
- Reports individual product performance.
- Monitors inventory usage.
- Processes credit cards immediately.
- Eliminates math errors and minimizes over/under-rings.
- Controls discounting.
- Ends errors caused by poor handwriting.
- Highlights possible theft of money and inventory.
- Records employee timekeeping.
- Tallies employee sales and performance.

General Retail POS Systems

There are literally hundreds of POS systems for general retail use. It can be confusing and an error can be costly. Don't rely on the recommendations of a computer consultant, as they aren't typically experienced number crunchers. *Your accountant should be able to provide you with recommendations.* Your local restaurant supply distributor will also offer POS systems or can provide you with local references.

In selecting a POS system and other accounting software, look for one with convenient training and support and fast *local* repair service. Research the software company thoroughly – for history of performance and long-term stability – as you'll need them long after the sale!

Choose your software first

Find the one with the features and reports that best suits your specific business. Then find the right hardware (cash register, computer, printers, etc.) to operate it.

Words of Caution: Custom programmed software that works only with the vendor's hardware means that you are tied to that company even if you are unhappy with their product or service.

The manufacturers' representative can provide you with sample reports and local references. CPA Online has a free research site on locating accounting software at **www.findaccountingsoftware.com** where you can also research POS systems.

Below are just a few general-purpose POS systems for retailers.

- QuickBooks Point-of-Sale Software (**www.intuit.com**) integrates with your QuickBooks for Retailers accounting software and offers a variety of merchant services including credit card processing.

- Microsoft Business Solutions (**www.microsoft.com/BusinessSolutions**) offers a variety of POS and credit card systems along with accounting.

- The Retail Solution offers general retail and restaurant-specific POS systems (**www.nwns.com**).

Bakery-Specific Systems

Bakery-specific computer systems are available for retail POS activities. These can be stand-alone POS systems or customized systems that integrate with production, inventory and accounting software modules. Below you'll find a variety of manufacturers and resources for food service or bakery-specific computer systems and both Windows- and Mac-based business software that can handle production, inventory, e-commerce and retail sales processes.

- ShopKeep™ - custom bakery software – **www.shopkeep.com**.

- Twin Peaks Software – wholesale distribution, retail POS, inventory and e-commerce software – **www.twinpeaks.net/twinpeaks/home.asp**.

- Mixing Bench software – product formula software – **www.rivik.com**.

- Nexcor Technologies – sanitation and maintenance software for bakeries – **www.nexcor.com**.

Food Service Software

The food service software packages below are what we refer to as "stand alone," as they are not part of a POS system. Although not bakery-specific, your bakery/deli operations may benefit from one of these.

ChefTec

ChefTec is an integrated software program with recipe and menu costing, inventory control and nutritional analysis.

Nutritional Analysis: Pre-loaded with USDA information. Add your own items. Calculate nutritional value for recipes and menus. Provide accurate, legal information on "low-fat", "low salt," etc. Print out "Nutrition Facts" labels. The nutritional-analysis module will give you a quick and accurate analysis of nutritional values for up to 5,000 most-commonly used ingredients.

- Add your own specialty items.
- Calculate nutritional values for your recipes and menu items.
- See at a glance which menu items are low fat, low-calorie, etc.
- Print a "Nutrition Facts" label.

Inventory Control: Preloaded inventory list of 1,900 commonly used ingredients with unlimited capacity for adding additional ingredients. Import purchases from on-line vendor's ordering systems. Track fluctuating food costs. Compare vendor pricing. See impact of price increases on recipes. Automate ordering with par values. Use handheld devices for inventory. Generate custom reports. The inventory control module allows you to track rising food costs automatically.

- Compare vendor pricing at the touch of a button, from purchases or bids.
- Enter invoices quickly using the "Auto-Populate" feature.
- Generate customized reports on purchases, price variances, bids and credits. Speeds up physical inventory, ordering and maintenance of par levels. Lists ingredients in different languages (Spanish, French, German and others).

NutraCoster

NutraCoster calculates the product cost (including labor, packaging and overhead) and nutritional content for any size batch of food. Ingredient database with nutritional information for approximately 6,000 ingredients (USDA Handbook 8). Include unlimited number of ingredients and process stats. Print camera-ready "Nutrition Facts" labels that comply with the requirements of the Nutrition Labeling and Education Act (NLEA).

The "Overhead Calculator" allows you to factor overhead costs into your cost of goods sold. No more "rules of thumb" or "fudge factors." Nutritional analysis now accounts for nutrient changes during processing, such as water lost during baking, water gained during boiling, fat lost during broiling, fat gained during frying, or other changes in nutritional value. Print multiple "Nutrition Facts" labels per page. Include ingredient listings with Nutrition Facts. Supports unlimited simultaneous users.

NutraCoster also offers additional libraries with nutritional information for specific brand-name ingredients. An accompanying *StockCoster* Inventory Module handles inventory control, production and vendor data and quotes.

MenuPro

MenuPro allows you to quickly create your own professional menus at a fraction of the cost of print-shop menus. Whether you need "Daily Specials" or take-out menu, MenuPro gives you quick, top-quality designs and artwork without the expense or hassle of using a graphic artist or desktop publisher.

Employee Schedule Partner

Employee Schedule Partner is a complete software package for employee scheduling. Point and click: make a schedule without touching the keyboard. Click a button and the software will fill your schedule with employees automatically. Click a button to replace absent employees, and a list of available employees' with phone numbers will appear. The on-line coach will give helpful hints to new users. Accommodates an unlimited number of employees and positions. You can manually override selections at any time and track employees' availability restrictions. Schedule employees to work multiple shifts per day, and track payroll and hourly schedule totals for easy budget management. Schedules can begin on any day of the week. Track stations as well as positions. Specify maximum hours per day, days per week and shifts per day for each employee. Lock any employee into a scheduled shift so the program will not move them when juggling the schedule. Save old schedules for reference when needed.

Employee Time Clock Partner

This is the complete employee time clock software package. It is very powerful yet simple to use. Automatically clock in and out (just enter your employee number and you are clocked in or out). Employees can view their time cards to verify information. Password protected so only management may edit time card information. Even calculates overtime both daily and weekly. Management can assign Employee ID number or PIN (personal identification number).

Accounting Software

Computer programs such as QuickBooks® (**www.quickbooks.com**) or sage® (**na. sage.com/us/sage-50-accounting**) are a solid choice for in-house bookkeeping. These programs are inexpensive, easy to use and will save time, money and countless errors.

Other Computer Uses

- Improve communications – websites, email, blogs, social media
- Promote your business – brochures, flyers
- Become your own "printer" – business cards, menus, discount coupons
- Attract customers – free Wi-Fi
- Recognize customers – photo displays, bulletin boards

Chapter 14

Public Areas of Your Bakery

irst impressions are important to have a successful business. Customers will judge you based upon your public areas. Your service counters, waiting area and dining room must be:

- **Attractive** – A visually stimulating place to shop or linger.
- **Clean** – Sanitation is a critical health issue. Since most customers never see your kitchen, they will judge you by what they do see.
- **Efficient** – Good service depends upon an efficient workplace.
- **Organized** – We live in a "fast food world." A well-organized self-service and work area will speed up your service.
- **Inviting** – A combination of sights, smells and personality that encourages people to return.

From your front door to your bathrooms, you can create a great impression by paying attention to details.

Creating a Design Focal Point

Your bakery should have a design concept that accentuates your products and a focal point from which to develop a brand. Here are some creative ideas to help you select and develop your bakery's *design focal point*.

- **Choose a design *focal point* that reflects your vision and build upon it.**
  - o   For example, if your focus is hand kneaded bread, incorporate it in your...
    - • Bakery name – Baker's Hands.
    - • Logo, signs – Strong hands with wheat.
    - • Uniforms – Natural colors of grain.
    - • Decorative accents – strong contrasting colors.

- **Sometimes your *design focal point* has no point at all.** Design for a feeling using elements make you think - elegant, cozy, warm, safe, wealthy, rested, young, spoiled.

- **Plan for the future.** Retail décor has a 5-7 year lifespan. Concentrate on versatile and neutral foundation elements (floors, lights, ceilings, fixtures) that won't need to be tossed when you redecorate.

- **Spend more on quality basics and save on the decorative touches.** Unlike retail stores where everything is part of the total look, using a *design focal point* allows you to accessorize without breaking the budget.

- **Don't over "design."** Use blank spaces (bare walls, neutral floors) to ease the eye.

Outdoor Areas

Your bakery could benefit from outdoor seating and dining areas. If you have significant foot traffic in the area, you'll encourage people to pick up a snack and beverage while resting. When considering an outdoor area, be certain to:

- **Review zoning regulations** for possible restrictions.

- **Strategically place plants, trees and decorative accents** to obscure unattractive views, shelter customers from the wind and soften noise levels.

- **Provide ample lighting to prevent slip and fall injuries.**

- **Warm chilly evenings with portable gas heaters, fireplaces and fire pits.**

- **Tame Mother Nature and protect customers** from sunburns or sudden showers with well-secured umbrellas, patio covers, pergolas, and awnings.

- **Use tables that are easy to clean and hold up to the weather.** Vinyl or recycled plastic can be scrubbed, hosed and are sliver free.

First Impressions

An attractive "front door" is a powerful welcome and invitation to dine. A well-planned exterior invites passersby to stop, guides customers to well-lit entries and previews the good things to come.

Here are some ideas for problem solving, mood setting and marketing from street to door.

Parking

- **Review the ease of access, traffic flow and available parking when selecting a location.** Don't forget to check local regulations and ADA requirements regarding accessibility for disabled customers.

- **Offer prominently located bike racks or free parking validation.**

- **Provide visible signage** to "steer" customers to your front door. Waiting -

- **Decide on whether you'll include exterior seating.** Long lines can be sign of a great hotspot or a signal that customers should go elsewhere for a *quick* meal.

- **Look for exterior seating materials that are easy to clean,** drain well, stay reasonably cool to the touch, won't snag or stain clothes or shed slivers.

- **Select chairs, benches or low "walls" that can be secured for stability and reflect your interior style.**

Mood Setting

- **Help differentiate your bakery** from the monochromatic; industrial looks of leased spaces in malls, strip centers and office buildings with colorful awnings, fresh flower boxes, attractive murals and signs.

- **Add music**, piped through exterior speakers, to set the mood and stimulate appetites.

- **Post menus and add "daily special" signage.** These are great marketing tools to lure in potential customers passing by.

- **Greet customers with positive smells.** Fill your landscape with colorful and fragrant flowers, place freshly baked goods near exterior vents, and use pleasant smelling cleaning products in entryways.

Window Displays

Window dressing is another creative way to make your bakery an inviting place to shop. With some basic carpentry tools, a bolt of fabric and some innovative uses for ordinary materials, you can make window-shopping a powerful lure for passers-by.

Don't let "stale" bread or fake food get dusty. Change your display often or people will quickly ignore its persuasive power. Bakery goods are typically shades of brown to punch up the presentation with vibrant colors, decorative plates and attractive signs.

If you have a broad window ledge, create a diorama by arranging cans or boxes of different sizes and cover them with draping fabric. This multi-level display will make each piece of food stand out. If you don't want people to remove items from the display – create a short barrier that won't block the view. In addition, don't forget to give customers an attractive inside view.

Two-sided freestanding displays can be created or purchased to add drama to a window and provide valuable shelving inside. If your bakery décor has a distinct style or theme, incorporate that into your displays and don't forget to promote holidays and major community events.

Visit your local craft stores, silk flower stores and commercial decorative supply outlets for ideas you can customize for your retail store.

Add a Touch of Green

Greenery and flowering plants inside and outside your bakery are great decorative touches that add life and vitality to an environment. Interior plants can even help filter the air and provide additional oxygen. Attractive landscaping can brighten a dull environment, hide unattractive exterior elements (such as garbage areas, adjacent buildings or freeways), create an entryway focal point, or provide a cool area to wait.

When incorporating plants into your design and décor, remember they require adequate light, moisture and accessibility. Below are some plant décor and landscaping hints for those without a green thumb.

- **Hire a plant maintenance or landscape firm to keep everything trimmed, feed and fresh looking.** Dead or spindly plants and yellowed leaves lying about are unappetizing signals for diners. Rotate interior plants regularly. Make certain the plants look good year around or can be inexpensively removed and replaced with more seasonally suitable choices.

- **Go faux!** Silk plants may be a better choice under some conditions. Select top quality artificial plants with flowers and leaves in natural colors. Bright blue blossoms may coordinate with your décor but they aren't frequently found in nature.

- **Create portable flower gardens** by using colorful pots, planters, hanging baskets, barrels or even old kitchen items. These can be great ways to soften hardscapes.

- **Avoid plants that are unpleasant smelling, have prickly needles, leaves or thorns or have toxic leaves/flowers/berries.**

- **Select the right plant for the space.** Consider light and watering requirements and full-growth size. Don't let vines dangle in someone's face or let upright plants interrupt an attractive dining room view.

- **Use full-spectrum lights for healthier indoor plants.**

Covering Your Floors

Although people might not gush about your flooring, it certainly influences their overall impressions of your bakery. In a busy environment with heavy foot traffic, flooring choices have lasting consequences and can really overwhelm your design budget.

Here are a few other issues you should consider when selecting flooring materials.

- **Choose commercial grade** whenever possible as anticipated usage and lifespan are typically much greater.
    o   **Check all manufacturers' warranties** for coverage in commercial applications.

- **Select materials for public areas that...**
    o   Won't show scuff marks easily
    o   Can handle chairs being dragged across or equipment being wheeled
    o   Won't be dented by high-heels.
    o   Have a medium colored pattern to hide spots, crumbs and dirt between cleanings.

- **Compare hardwood flooring with modern vinyl or acrylic-infused look-alikes.**

- o   Remember wood can be sanded and refinished easily while the look-alike would need to be replaced.
- o   Select the more expensive strip vinyl flooring for a longer life expectancy than other vinyl products. Replacing small damaged areas is an added benefit.

- **Ask your architect about the great ways concrete is being used in commercial buildings.** No we're not talking about floor that looks like a driveway. New processes and color techniques make this an attractive and durable choice.

- **Avoid dark, high-gloss flooring,** which show dirt easily can appear wavy and magnifies any substructure imperfections.

- **Make certain all flooring is…**
  - o   Easy to maintain.
  - o   Can handle chemical exposure.
  - o   Slip-resistant in wet and dry conditions.
  - o   Meets sanitation code in food prep areas.

Ceilings

Ceilings are often overlooked when designing and decorating a bakery. Customers actually do notice attractive colors, artistic displays and great lighting along with all the dust, cobwebs, stains and ugly ceiling materials. An attractive and clean ceiling tells customers you value cleanliness and care about them.

When choosing ceiling materials, look for sound-deadening materials that are easy to clean and secure tightly to beams, sheetrock or suspension hardware.

Below are some things you should know about choosing ceiling materials, designing unique ceilings and maintaining ceilings.

- **Use moisture-proof, mildew-resistant materials** that meet your local sanitary standards on ceilings in high moisture areas (food prep, dishwashing and restroom areas).

- **Transform ceilings with wallpaper, wood paneling, fabric or other suspended treatments.** Just be certain all materials are fire-resistant and meet code.

- **Use exposed beams, pipes and vents as great color accents and high tech art pieces.** Make certain paint and other treatments are fireproof and heat-resistant for heating and steam pipes and waterproof for water pipes.

- **Reflect more light and make the room feel larger with lighter colored ceilings.** Remember lighter colored ceilings will also show venting-related dirt stains. Your local Health Departments may require light ceilings in work areas to aid inspectors.

- **Make certain your HVAC is properly vented and well maintained to eliminate ceiling stains.** Lack of maintenance isn't just unsightly; it wastes electricity.

- **Incorporate skylights, light tubes and windows to bring in more natural lights.** Make certain these can be cleaned easily at least once a quarter.

- **Think of your ceiling as another wall to be decorated.** Tin ceilings, faux painting techniques, mirrors, posters, faux beams, decorative molding and fabric are all potential ways to add drama, carry out a theme or enhance a peaceful environment.

Let There Be Light

Good lightening creates a mood, enhances décor, makes it easier and safer to work, and makes you and food look better.

When considering how to light your display cases and service counters, here are some things you will have to consider:

1. Level of natural light and seasonal changes which affect it.
2. Activities within the room – work areas, walkways, tables, waiting areas.
3. Ambiance you wish to create – bright and simulating or soft and romantic.
4. Artistic and creative uses – the use of light and shadows to accent attractive features or mask less attractive areas.

Lighting effects can be obtained through wall sconces, fiber optics, chandeliers, track lighting, table lamps, directional spotlights, and even candles.

- **Incorporate indirect lighting.** Well-placed wall sconces add light without the glare.

- **Use color-accurate lighting to enhance the food's appearance.**
  Incandescent lighting has a warmer, yellow-orange cast; fluorescent lighting
  produces a blue-green cast. *A real appetite deadener!* Halogen lights are
  closest to true white light.

- **Explore full-spectrum lighting** (which reportedly make people feel
  healthier) for work areas.

- **Hire a lighting designer.** This lighting expert can help you upgrade existing
  lighting for appearance and energy savings or design a complete new look.

- Visit **GE Lighting** online at **www.gelighting.com/LightingWeb/na/** for
  design, product selection and energy audits.

Colors That Complement

Most food in a bakery is brown or beige. That leaves you with some great opportunities to enhance your bakery and your food by adding decorative colors. Scientists have proven that people are affected by the colors surrounding them. Why not incorporate one or two to create the right mood for your bakery?

Yellow	Sunlight, cheerful, vitality. Many designers believe every room should have a dash of yellow. Stay away from greenish yellows.
Red	Intensity, passion, stimulates appetites. Use boldly or as an accent.
Blue	Cool, clean, and refreshing. Blue should be used away from food as it isn't complementary.
Green	Well-being, nature, fresh and light. Beware - can also make people and food look off-color.
Gold	Wealth and power. Warms up other colors and brightens dark wood.
Neutrals	Masculine – darker browns. Feminine – lighter terra cotta shades. Rosy hues make food and people more attractive. Rarely go out of style and provide a background for bold color accents.
White	Clean, fresh and new. Can be a good foundation color but beware it can also signal institutional, bland, ordinary. Can create glare and eyestrain.
Black	Death and mourning. However, used properly black can add elegance and style. Black and white is a classic look. Avoid as a background

color and don't forget that black can show fingerprints and can be difficult to keep looking clean.

A Little Artwork

Finding the right artwork for your bakery may mean that you have to commission an artist to create the perfect piece. Hiring an artist isn't like hiring any other professional. Reputations, egos and biases can mean head bumping and unhappy outcomes. Also what constitutes "art" is in the eye of the beholder and can frequently be difficult to define. Working closely with an artist during the conceptualization stage is critical.

Your Restrooms

Restrooms may be the smallest rooms in the house but they are important ones for guests. An ample, clean restroom speaks loudly about how you value cleanliness and are considerate their needs.

Plumbing and health department standards vary widely across the U.S. Also, the *Americans with Disabilities Act* (ADA) governs accessibility issues for all public places. Be certain that you comply; as inadequate restrooms can keep you from opening.

Here are some practical and creative ideas on designing and decorating restrooms.

- **Remember your customer's physical needs.** Provide sinks, dryers and dispensers at levels appropriate for children and wheelchair-bound patrons. For ADA advice, call the U.S. Dept. of Justice at 800-514-0301.

- **Select materials wear well, won't show dirt and can handle strong cleaners.** Ceramic tile is great but be aware of grout discoloration.

- **Avoid public unisex rooms whenever possible.** This may be a practical solution and "politically correct" but most people still feel uncomfortable. Also some jurisdictions require separate facilities for men and women. The exception to this is an oversized "Parent & Child" restroom.

- **Provide separate facilities for staff if possible.**

Create a Presentation

Displays are your best presentation tool. Whether they are self-service displays or rotating refrigerated cases, a great looking display can boost your bottom line by:

1. Increasing impulse buys.

2. Showing off your decorator's talent.
3. Reducing labor costs through self-serve.
4. Preserving perishables.
5. Saving energy.

Your display style should connect with your customer. Most bakers prefer to have their shelves brimming with freshly baked items. Empty trays mean lost sales. However, a minimalist approach where a few muffins or individual pots de crème are artfully displayed on a platter can be appealing – to the right audience. Your personal presentation style and your customer's needs should be complementary – if not you won't be happy and neither will they.

Displays aren't just for storage – they are for merchandising. Think presentation, presentation, presentation. Decorate displays for the holidays; add orchids in stem vials; incorporate antiques; use collectible plates or large charges in food-safe gold with paper doilies; create multiple level vignettes; and angle platters towards the customer.

Once you get your creative juices flowing, you'll find that merchandising your food can be a great creative outlet that boosts sales! Remember that all food displays must meet local health ordinances so if in doubt – ask first.

Display Cases

Display cases, also called showcases, can be horizontal or vertical displays capable of holding dry goods or refrigerated items. Knowing your product mix and front of the store layout will help you decide on the best cases for your needs.

When selecting a display case, look for:

- Easy to clean units with removable components and no hard to reach areas.
- Adjustable shelving that can accommodate the tallest cake or multilevel displays.
- Energy efficient refrigerated models. Compare energy ratings. To lower your utility bills:
  o Use insulated night covers on display cases.
  o Clean condenser coils, fans and gaskets weekly.
  o Inspect rubber sealant and gaskets regularly – replace when dry or cracked.
  o Remove "unnecessary" internal case lights if you have ample ambient light.

o   Check and calibrate the thermometer regularly.

o   Don't overload refrigerated cases.

o   Select the higher energy efficiency rating (EER). The greater the cooling capacity for each kilowatt-hour of energy input, the greater the efficiency of the system.

o   Avoid older used showcases – your initial savings probably won't offset the extra energy cost.

When choosing a display, think about how a grocery store merchandises cereal – the most expensive "adult" products are at the top (within an adult's reach), the desirable high profit "kids" products are stocked in the middle of the shelves (impulse items for kids) and the lower profit bulk and generic brands are on the lowest shelves. Determine your low, average and high profit items and display them accordingly. High traffic items such as bagels or donuts in the morning and loaves of bread in the afternoon can be stocked on an eye-level shelf and the "off schedule" items moved to a less prominent position.

Having ample displays is critical so carefully chart out your retail area to maximize the space with undercounter displays, wall-sized displays, freestanding vertical merchandisers and movable units for temporary promotions. When laying the room out, think about how your ideal customer will enter and approach each display. Will they sweep the room looking for just the right loaf of bread or cake? Will they walk directly up to the counter to order?

Many bakery owners have discovered that non-bakery and non-food items can be excellent profit centers. When selecting displays, think about the need to hold stacked coffee mugs, display birthday candles and hats or boxed tea sets. You can create an area away from the cash register to display these or merchandise them by placing non-food items next to the companion food items (birthday candles next to the birthday cakes).

Counter Top Displays

Eye-level displays on your counter can be quite effective in merchandise new food items or hot enticements like fresh cookies. Just don't overwhelm the active work areas and cash register. People don't like to feel like people are pressing in around them when they are opening their purse or wallet.

These displays can be temporary cardboard merchandise displays provided by the supplier, attractive chrome and glass units, unique bowls or stands – just about anything that makes it easy for customers to make a last minute choice just before being rung up.

Self-Service Displays

Self-service displays and beverage dispensers save labor and can help move traffic in and out quickly. Self-service units should be placed out of the main traffic pattern so browsers don't disrupt other customers. Limit your self-serve items to packaged goods such as loaves of bread or cartons of juice. As "bulk" displays are considered to be unsanitary.

If you have "help yourself" beverage dispensers, you'll need to have someone assigned to keeping the area clean and tidy. Contact points (scoops, nozzles) should be cleaned or changed regularly. Your coffee and soda vendor typically will provide you with free or economical dispensers when you feature their products.

Self-serve displays can be:

- Functional – traditional sliding or swinging glass doors.
- Elegant –European-style wood racks.
- Temporary – freestanding cardboard displays.
- Classic – old-fashioned tubs with ice holding bottles of water.
- Creative – combine boxes, shelves, stands, wagons, strollers and décor for innovative displays

Ice Cream Freezers/Milk Dispensers

For in-store dining, you might want to consider an ice cream freezer *(pie ala mode anyone?)*, milk dispensers *(just the thing to drink with warm cookies)*. Many other companies offer similar contracts for the use of equipment when you guarantee to purchase their product exclusively. Ice cream freezers), is available. Your sales representative will have all the information about the equipment available.

Compare the cost of these with buying tubs of ice cream and individual serving milk cartoons. Don't over invest in this type of equipment – especially if the majority of your traffic is take-out.

These arrangements may be very beneficial for small bakeries that have limited capital. Whatever your financial situation is, do a thorough, careful investigation into the terms of the service contract. In some cases, the price of the product may be so high it would be better off purchasing equipment. You should always compare competitive prices of the products after several months of operation. The free equipment may not justify the total cost of the product. Should you decide to sign the contract and use the free equipment, remember that you may be locked in for a long time. Be sure to request new equipment from the sales representative.

All the freezers and dispensers left in the open should have some sort of locking device on them to protect against tampering.

Front-of-the-House Work Areas

Realistically, not all food prep and service work can be accomplished behind closed doors. Here are some different types of work areas you might need in the front-of-the-house. Even if your sole workstation is a long service counter, divide the work area into specific activities. A typical bakery's front-of-the-house will need to:

- Display prepared food – cakes, bread, and rolls.
- Package food for travel – box donuts, slice and bag bread.
- Take orders – fill orders immediately and write up special orders.
- Ring up sales and make change – operate a cash register.
- Prepare food and beverage – make a cappuccino
- Finish food items – warm a sweet roll.
- Store packaging, serving ware, clean-up materials and cash

Here are some helpful suggestions on designing and implementing front-of-the-house work areas.

- Make them attractive without sacrificing function.
- Remember cleanliness and order are required when service personnel prepares food in full view of customers.
- Don't forget to build in floor drains, use scuff resistant baseboards and add casters to equipment that must be moved for cleaning.
- Reduce lifting and carrying with mobile carts and rolling waste receptacles.
- Use properly aimed task lighting to avoid glare while allowing staff full visibility of the work surface.
- Use anti-fatigue mats and non-slip flooring.
- Design work areas to minimize stooping, reaching and lifting.
- Separate "wet" and "dry" tasks to avoid damage, food contamination and electrical accidents.
- Incorporate hand and/or utility sinks whenever possible to save steps and promote cleanliness.
- Provide ample counter space below pass-throughs to verify orders and fill trays.

Shelving

Shelving, as opposed to displays, is primarily storage for activities that take place behind the counter. To improve your counter staff's efficiency you'll need to store a variety of tools, equipment, supplies and food items within easy reach.

Out of the customers' view, storage units (under the counter or with doors) should be where you keep "unattractive" items such as cash register tape and cleaning supplies. Open shelving facing the customer can be artfully arranged for appearance and functionality. You'll need to balance functionality with appearance and orderliness.

Senses Check

Regularly walk up to your front door, look around, enter and take in the sights, sounds and smells.

- What catches your eye immediately? First impressions do count!
- Do you feel like this is a clean, well cared-for establishment?
- Does it feel comfortable and inviting?
- Are you looking at half-filled cans, dripping syrup bottles or ugly trashcans?
- Do you detect the harsh smell of cleanser or the aroma of bread baking?
- Can you hear the counter person over the clanging of pots and pans?

Look for inconsistencies in your message of quality and correct them immediately. Don't forget to assign someone to make certain that all public areas are kept tidy and inviting throughout the day.

Have fun with your merchandising and decorating – you'll enjoy it and your customers will appreciate it!

Chapter 15

Dedicated Work Areas

• •

Beyond your kitchen and public areas, there are other rooms or work areas that directly affect your productivity. How you handle incoming deliveries, store the food and supplies, deal with and deal with the waste can even affect your profitability?

Delivery Areas

Whether your facility has a loading dock or just back access door, it is important that you have the equipment and procedures to accept, log in and prepare for storage. When selecting vendors, inquire about their delivery service. How do they package heavy bulk items? Will they leave racked goods? Will they just drop everything at the door or will they move bags of sugar and flour to storage areas for you?

You may not require a forklift for palletized deliveries, but there are other loaders and lifters that can speed up the process and minimize employee injury. Your local food service supply company and food service equipment vendors can help you select the best

equipment for the task. Below are some recommendations on outfitting your delivery area:

- Don't forget to provide employees with protective gear such as gloves and heavy-lifting belts.

- Having computer access in this area can speed check-in. Packing slips can be quickly compared with orders, accepted and signed off. However, a clipboard with copies of orders can suffice.

- Create a procedural manual. Each vendor will have a written damaged goods and return policy. Keep copes in the binder to support employee training.

Your employees should understand proper delivery acceptance procedures. Below are some procedures they must know to catch your ordering errors and vendor shipment errors.

- All visibly damaged merchandise must be noted on packing slips or bills of lading (or depending upon the vendor's recommendations – refused). This helps the vendor file third-party carrier claims.

- Hidden damage should be noted upon the packing slip and management advised immediately so claims can be filed.

- Products should be inspected for package damage; signs of pests and excess debris; and mishandling.

- Over- or shortages should likewise be noted on the packing slips or bills of lading.

- Employees must sanitize their hands and remove soiled aprons to avoid potential cross-contamination with opened goods.

- Realize that the delivery person may not be at fault. Take your complaints to the vendor's customer service department or your sales representative.

Pre-Storage Activities

Some items such as fresh produce may require pre-storage prep to length storage times and ensure its quality.

Items being prepared for freezing may require prechilling to speed the process and preserve the food's freshness. Breaking down items into convenient units can also speed production times and minimize waste.

Storage

You'll need to store raw ingredients, supplies and finished goods. You'll need refrigerated storage, freezer storage and dry storage equipment and areas. You'll have long-term and short-term storage needs. Before you can determine your storage needs, you'll need to develop your product mix and menu. Knowing the ingredients you'll be purchasing will help you make better choices for storage.

Increase productivity by creating three types of storage.

1. **Active** – accessed repeatedly throughout the day. Located closest to the active work area.

2. **Back up** – refill (bulk) items for active areas and items used occasionally during one week. Located further from the active work area but easily accessible.

3. **Long term** – non-perishable special use and seasonal items. Uses out-of-reach, back of the building, under stairs and less accessible areas.

Dry Storage

Your dry storage should be conveniently located for deliveries and for employee access. Access doors should be ample enough for lift equipment and racks capable of holding 100-pound bags of flour. All shelving should be at least 6-inches from the floor to deter pests.

Below are some practical ideas on creating useful storage areas.

- Make storage cabinets in public areas attractive, a part of the décor and from materials that clean easily.

- Create separate (but convenient) storage for chemical cleaners and other hazardous materials. Check your local regulations regarding hazardous material storage.

- Evaluate all storage for potential cross-contamination issues. This includes chemicals, foodstuff and handling methods.

- Incorporate easily movable or "sectional" storage whenever possible to maximize layout flexibility.

- Protect employees from injury by placing heavy items closest to waist height as possible. Provide sturdy stepstools, ladders and rolling carts nearby. Except for rarely accessed areas, keep shelving shallow enough for easy reach.

Refrigeration

Commercial refrigeration is a must. Consumer models do not provide as even a temperature or ample room. Look for a model that provides you with sufficient space to hold mixing bowls to retard dough (keep from rising), firm up whipping cream and keep frosting stable.

Freezer

You may not have to invest in a freezer if your products are made fresh daily. However, you may find that quick-frozen fruits or vegetables are a better choice than inferior off-season goods for some uses. As with refrigerators, allow ample space so food will chill faster.

Waste & Recycling

Handling food, oil/grease and solid waste takes time, effort and money. Waste management and reduction are not only cost-effective; they are wise environmental and business choices. Here are some helpful ideas on managing and reducing your bakery waste and disposal costs.

- **Purchase a waste disposal unit.** Look for stainless steel units with automatic reversal controls. Select units with ample horsepower and rotor size to handle your typical of food waste. Invest in long-term performance when comparison-shopping.

- **Build a recycling center and establish a usage program.**
  - o   Include recycling equipment in your back-of-the-house layout.
    - Contact grease/meat waste rendering companies about pickup.
    - Make it easy for employees to comply by incorporating sorting bins, conveniently located waste receptacles. Install color-coded recycling containers on wheels.

- Be prepared to handle large quantities of paper, cardboard, plastics, glass, metal cans and food waste.
  - Cardboard balers for example can pay for themselves through reduced hauling costs.

- **Invest in a commercial grade trash compactor.** Even with the most aggressive recycled program, you will still have trash. A compactor will pay for itself quickly by maximizing bin use and hauling fees.

- **Talk with local recycling companies and government waste management agencies.** Recycling companies can handle grease/oil waste; pick up glass and cardboard. There are even companies that specialize in bakery waste. Your city or county can assist with developing waste reduction programs.

Your Office

Every business needs an office. If you choose to handle your bookwork at home, you should still have an in-house office. That can be an entire room or an isolated work area. You'll need ample lighting (natural or artificial), filing cabinets (including locking ones) and a suitable desk and chair. Your computer could be a single workstation or part of your bakery's computerized network.

- Create an organized and productive work area so others, in your absence, can locate, file and process paperwork.

- Store confidential documents in locked file cabinets.

- Keep copies of all policy and procedural manuals available for reference.

As a baker, you probably feel most comfortable in the kitchen; so don't forget to make your office comfortable – physically and emotionally. Important work goes on here so create a stimulating and productive environment.

Chapter 16

Creating a Safe and Productive Business and Work Environment

• •

he investments you make in creating a work environment that protects your employees and your community will reward your business with increased productivity and savings. The International Labour Organization details potential bakery worker safety issues at **www.ilo.org** (use keyword searches "baker" and "bakery").

Good Ergonomics

Ergonomics is the study and engineering of human physical interaction with spaces and objects during activities. A prep area that requires workers to repeated stretch across to reach ingredients or a broiler unit that only very tall workers can safely reach is "poor" ergonomics.

Good ergonomics in a restaurant can positively affect your employee's physical well-being, safety, productivity and comfort. Good ergonomics, such as well-fitting tables and comfortable chairs, can also enhance your diner's experience.

Here are some valuable tips to help you "engineer" your restaurant to work well with people. Additional in-depth information can be found in chapters covering with specific issues such as lighting, equipment, work-flow and traffic patterns.

- **Create mini-work stations where all necessary food, utensils and prep space are close at hand.**

- **Eliminate excessive bending, lifting, and reaching while encouraging proper prep and storage procedures.**

- **Provide stools or chairs to give backs and feet a rest,** if the work being done doesn't require standing,

- **Make certain your tools and equipment weren't designed for only men.** Although more and more women are donning toques, tools and equipment haven't necessarily been redesigned to accommodate their shorter frames, smaller hands or differing physical characteristics.

- **Provide stable, heavy-duty work ladders for accessing top shelves and deep storage units.**

- **Purchase a supply of important tools and utensils for left handed employees.**

- **Think about how employees, customers and vendors will interact with your facility.** Does the facility or equipment make it easier or more difficult to do their job or enjoy their shopping experience?

- **Choose fixtures and equipment that can be easily moved when needed from work area to work area.**

It's Good To Be Green

In building and/or renovating your bakery facility, you have an opportunity to incorporate green materials and designs. Here are some resources to learn more about creating a less wasteful and more ecologically productive business.

- **Learn more** about the green construction movement and how it can improve your bottom line at:
  o   Green Architect at **www.archrecord.com/GREEN/GREEN.ASP**.
  o   U.S. Department of Energy at **www.eren.doe.gov/buildings/ commercial_roadmap**.

- **Look for grants, low-interest loans and funding for green construction.**
  - o Funding Green Buildings at **www.fundinggreenbuildings.com** are funding consultants specializing in green building.
  - o ShoreBank Pacific offers eco-loans for Pacific Northwest businesses. Info at **www.eco-bank.com**.
  - o Fat Earth at **www.fatearth.com** offers a financing resource directory and other information on paying for energy-saving and green construction and equipment.

- **Hire green building professionals.** A service directory can be found at Greenbuilder.com at **www.greenbuilder.com** or search Google by the keyword "green building."

- **Pick up a book on the subject.** As this is a rapidly changing industry, select the newest books available. Popular titles include:
  - o *Green Building Handbook: A Companion Guide to Building Products and Their Impact on the Environment* by Tom Wolley.
  - o *Green Building Materials: A Guide to Product Selection and Specification* by Ross Spiegel, Dru Meadows.

- **Require your architect and builder to comply with the EPA's Energy Star Design Target program.** Details at **http://www.energystar.gov/ buildings/service-providers/design/step-step-process/evaluate-target/ epa's-target-finder-calculator**.

- **Research sustainable and green resources** at Fat Earth at **www.fatearth. com**. Search for restaurant-specific info using the keyword "restaurant."

- **Use your green building efforts as a springboard for free local and regional publicity.**
  - o Register your building and efforts with the EPA's Energy Star Program.
  - o Contact your utility company(s) and explore joint PR opportunities.

Conserving Energy

Food service environmental systems and equipment require an enormous amount of energy. The National Restaurant Association reports potential profit increases from 4 to 4.5 percent of revenue simply by reducing energy consumption by 25 percent.

Energy-conscious construction methods, along with energy-efficient equipment and environmental systems, are wise investments. Although initial costs may be greater, your ROI could actually offset the **full-cost** of the equipment over its lifespan.

Below we've detailed some useful tips to help you create an energy-wise bakery.

- **Have your local utility company complete a free energy assessment on your existing facility.** They can monitor the efficiency of your walk-in freezer, cooler or oven to help you determine your actual energy cost per month for each. Use these figures to compute potential energy savings if you replaced the unit with a more efficient model.

- **Concentrate first on the major energy users.** HVAC (heating, ventilation, air conditioning) systems account for 30 to 50 percent of your annual electricity costs.

- **Don't overlook small ways to save** such as more efficient light bulbs, programmable thermostats, and plastic strips on walk-in refrigerators.

- **Visit www.epa.gov to learn about the EPA's Energy Star program** designed to help small businesses become more energy efficient. Restaurant-specific information is detailed here.

- **Seek out private and government lenders who specialize in financing energy-related improvements for the most favorable interest rates.**

- **Contact your utility companies about subsidies, rebates and financial incentives** for replacing inefficient equipment with approved models. Your tax accountant can also advise you on state or federal tax incentives.

- **Hire an energy consultant** to make recommendations on your design and equipment choices.

- **Remember you are investing in your bottom-line with long-term payoffs** for your restaurant and our environment. Energy costs have historically risen, so your savings factor could be even more significant in future years.

Quality Air

Healthy air, inside and out, is a business and moral concern that impacts restaurants legally and fiscally. "Poor air" contributes directly to employee absenteeism and unhappy customers. Many communities have rigid air emission and work environment regulations relating to proper ventilation, wood burning, grease and smoke. Unpleasant odors also contribute to "poor" air quality.

Fresh Indoor Air

Wood burning ovens, charbroilers, fryers and "sealed" buildings can all create unhealthy or unpleasant air conditions. In a bakery facility, flour dust is also of concern. Indoor air quality requires bringing sufficient outdoor air in, properly filtering the outdoor and recirculated air and directing airflow.

Below are several examples of helpful tips on improving indoor air quality.

- **Physically separate smoking and non-smoking dining areas** and/or direct airflow away from non-smoking tables. Ban employee smoking in the kitchen, dining room and bar.

- **Install a whole-building air cleaner/filtration system** that also reduces airborne particles and dust.

- **Check for radon, mold spores and biological dangers** when converting older or long vacant buildings.

- **Read what the EPA says about indoor air quality** at **www.epa.gov/iaq/ pubs/insidestory.html**.

- **Be aware of unhealthy emissions from carpeting, paint and cleaning products.** Sick Building Syndrome is explained at the National Safety Council site (**http://www.nsc.org/news_resources/Resources/ Documents/Sick_Building_Syndrome.pdf**).

- **Hire a HVAC contractor or engineer with bakery experience.** Hire contractors to install new systems or maintain existing systems and an engineer to design and specify systems.

Outdoor Air Quality Issues

Ovens, fryers, broilers and other cooking equipment all emit particulates, gases, grease and orders that are regulated by local, state and federal environmental standards. Local and state standards vary greatly and in some instances, the Federal Air Quality standards may supersede these. Close attention should be paid to these regulations, as penalties can be severe. Below are some helpful ideas on how to meet emission regulations.

- **Hire an industrial air-cleaning firm to install emission control systems** that handle grease, smoke, CO_2 and odors.

- **Inspect and repair all exterior vents, hoods and intake ducts on existing or older equipment.** Maintenance not only saves the air, it also saves you in energy costs. Dirty, inadequate or old systems are big electricity wasters.

- **Install a catalytic oxidizer that converts gases and smoke to water.** For more info, read the Products Finishing Magazine article at **www.pfonline. com/articles/010203.html**.

- **Contact your natural gas and electric utility company along with county or state Environmental or Health Department for air quality information, resources, and financial incentives.** The Southern California Air Quality Management District, with some the most stringent air quality regulations in the nation, offers comprehensive info at **www.aqmd.gov**.

- **Hire an air quality consultant** to assist you with more complex emission issues and stringent regulations.

- **Visit the Environmental Protection Agency's** Web site (**www.epa.gov**) for bakery specific regulations including the voluntary Bakery Partnership Program.

H_2O and Baking

In some communities, your tap water isn't very appetizing. Unsavory odors and harsh tastes along with hard water deposits are common issues. Unseen bacteria and pollutants may also be a concern. Poor water can adversely affect your food quality, damage equipment and tableware and disturb health-conscious shoppers.

- To learn more about water quality in baked goods, read
  - o *Bakery Food Manufacture and Quality: Water Control and Effects* by S. Cauvain and L. Young.
  - o *Liquid Assets for Your Bakery* (Publication # CD-41) at the North Carolina Cooperative Extension Service **www.bae.ncsu.edu**.

o   *Water Types, Affects and Solutions* at **www.bakingbusiness.com** (search "water").

Filtration systems act like a very fine sieve. Filters also attract and hold even smaller particles and dissolved molecules. Filters are rated by the particle size they remove. The smallest bacteria are one-half micron while visible "dirt" ranges to 40 microns. Activated carbon is a common filter medium.

Reverse osmosis forces incoming water under pressure through a filter membrane to separate undesirable elements. Your daily water usage determines the capacity required.

True purification methods use heat to distill water or add iodine to neutralize or eliminate containments. Distilled water may be too flat tasting and requires more energy to produce; while iodine may cause health problems and adds an undesirable taste.

Below are some additional tips on offering your customers clean, fresh and tasty water.

- **Determined what's in your water.** Your water department or independent testing company can provide your water's composition (solids, hardness, chlorine levels). Select equipment that addresses your water's "bad" elements. Besides safe and tasty water, your specific chemical and equipment needs for washing produce, equipment and dishware are affected by your water's composition.

- **If whole "house" systems aren't a cost-effective choice, incorporate smaller undercounter models** to provide clean water for prep sinks, beverage makers/dispensers and cooking water faucets.

- **Look for filtration systems that can be easily cleaned or use inexpensive replaceable filters.** Filters should be ample enough for a 6-month maintenance cycle.

- **Don't overlook clean, filtered water for icemakers.**

- **Choose systems that can reduce chlorine levels to approximately .5 parts per million for the best tasting coffee/tea.**

- **Sell high-profit sparkling and still bottled water.** Some customers prefer the "safety" and convenience of bottled water!

Waste Management

Handling food, oil/grease and solid waste takes time, effort and money. Waste management and reduction are not only cost-effective; they are wise environmental and business choices. Your efforts can actually become a marketing advantage. Customers really do appreciate knowing that their favorite restaurant cares enough to reduce, recycle and properly dispose of waste.

Here are some helpful ideas on managing and reducing your restaurant waste and disposal costs.

- **Check on local waste management requirements early in the project.** Municipal standards vary widely. Some jurisdictions may require retrofitting existing buildings when ownership changes.

- **Install a properly sized commercial grease trap and/or interceptor.** The Plumbing and Drainage Institute at 800-589-8956 or **www.pdionline.org** has comprehensive articles and member referral info. Grease containment equipment can run as much as $20,000 to install.

- **Research bioaugmentation.** Biological elements "digest" the fat and grease reducing the amount of waste your system must handle.
  - o Visit GE's Water & Power website at **http://www.gewater.com/ products/wastewater-bioaugmentation-products.html** for more info.
  - o Research local providers. This is "real" science but "quacks" are common. Carefully check references.

- **Purchase a waste disposal unit.** Look for stainless steel units with automatic reversal controls. Select units with ample horsepower and rotor size to handle your typical of food waste. Invest in long-term performance when comparison-shopping.

- **Build a recycling center and establish a usage program.**
  - o Include recycling equipment in your back-of-the-house layout.
    - Contact grease/meat waste rendering companies about pickup.

- **Make it easy for employees to comply** by incorporating sorting bins, conveniently located waste receptacles. Install color-coded recycling containers on wheels.

- **Be prepared to handle large quantities of paper, cardboard, plastics, glass, metal cans and food waste.**
  - o Cardboard balers – can pay for themselves through reduced hauling costs.

- **Invest in a commercial grade trash compactor.** Even with the most aggressive recycled program, you will still have trash. A compactor will pay for itself quickly by maximizing bin use and hauling fees.

Preventing Injuries

Your bakery employees have the same personal risks as any food service and manufacturing facility. When building your plant, buying your equipment, creating your workflow layout and establishing production procedures, you must be aware of the potential injury.

- **Slip and fall injuries** – Use rubber mats, non-skid floor surfaces, fast-draining floor drains and keep walkways clear.

- **Burns and cuts** – Look for equipment that doesn't require reaching over hot components. Buy high-temp silicone oven mitts and handle holders. Add pot-filler faucets by stoves. Make certain knives are sharp and never store sharp objects in a drawer.

- **Standing, stooping and reaching injuries** – Use soft rubber mats to stand on. Have hand-lifts, trolleys and rolling cart readily available. Store heavy items so they can be easily moved to these without lifting. Create storage and shelves within easy reach of shorter workers.

- **Repetitive stress injuries** – Review ergonomics of workstations, provide ample breaks and rotate staff on production lines.

- **Inhalation injuries** – Install proper ventilation systems and use particle masks when handling dry ingredients. The US Department of Labor, Occupational Safety and Health Administration (OSHA) and other government bodies worldwide have documented potential health risks from repeatedly inhaling flour dust. To learn more about this issue and equipment compliance, visit OSHA at **www.osha.gov**.

- **Eye injuries** – Train and retain employees on proper handling of chemicals. Have eyewash kit readily available.

As a retail operation, you'll also have to be diligent in protecting your customers from injury. Slip and fall injuries are the most common. Review your premises and implement preventative measures such as:

- Keeping the floors clear and clean. Make certain route to restroom is well lit.

- Using industrial non-slip floor mats and rotate them when they get wet.

- Keeping your entrance ice-, snow- and debris-free.

- Checking for slivers and sharp edges on counters and displays.

- Kid-proofing lower shelves by removing things that break easily or could be swallowed.

Chapter 17

The Essentials of Food Safety, HACCP and Sanitation Practices

Every bakery employee is responsible for preparing and serving quality and safe food products. Each employee must be thoroughly familiar with basic food safety and sanitation practices. This chapter will describe the fundamental methods and procedures that must be practiced in order to control food contamination, the spread of infectious diseases and personal safety practices. You'll also find information on customer sanitation issues such as trash receptacles and public restrooms.

You must provide employees with the training, knowledge and tools that will enable them to establish and practice proper food handling and sanitation procedures.

Aside from what is required by law, you must provide:

- *Training* including hands-on training sessions, refresher courses and manuals.

- *Personnel safety and hygiene supplies* including hairnets, uniforms, protective gloves, germicidal hand soaps, hand and nail brushes and first-aid kits.

- *Clean-up stations* including hand sinks at every work station, sanitary employee bathrooms and lockers, scrub brushes, gloves and disposable towels.

- *Equipment and storage supplies* such as labels for dating and rotation procedures, color-coded utensils, test kits and quality control standards.

Food preparation establishments may harbor all types of bacteria, bugs and animal pests. Bakeries attract these unwanted intruders by providing the three basic ingredients to sustain life: food, water and warmth. In order to eliminate contamination, you must make the living conditions unfavorable for these pests.

While your bakery may not be handling raw chicken or egg salad sandwiches, the information here applies to a variety of raw ingredients and finished food products that might be found in a food service establishment.

These safety principles are important to you, your business and your customers. Food industry experts believe that for every dollar you spend on food preparation safety, you'll save $10 on reduced waste costs and up to $100 for legal liability.

What Is HACCP?

Hazard Analysis of Critical Control Points (HACCP) is a system for monitoring the food preparation stages to reduce the risk of food-borne illness. HACCP focuses on how food flows through the process – from purchasing to serving. At each step in the food-preparation process, there are a  variety of potential hazards. HACCP provides managers with a framework for implementing control procedures for each hazard. It does this by identifying critical control points (or CCPs). These points are where bacteria or other harmful organisms may grow or food may become contaminated.

Why Use HACCP In Your Facility?

As a food purveyor, you are responsible for protecting your customers by serving safe and wholesome food. To accomplish this, you have to educate and motivate your employees to actively practice food safety procedures. You will need a systematic process for:

1. Identify¬ing potential hazards.

2. Implementing safety procedures.

3. Monitoring the success of your safety system on an ongoing basis.

HACCP is based on the principle – *if the raw ingredients are safe and the process is safe, then the finished product is safe.* Using HACCP, you can identify potentially hazardous foods and places in the food-preparation process where bacterial contamination, survival and growth can occur. Then you can take action to minimize the danger. Implementing HACCP involves seven key steps.

Step 1: Assess the Hazards

- Track each food from purchasing and receiving through serving and, if appropriate, reheating.

- Review your finished goods and menu items. Identify all potentially hazardous foods, as well as those foods that may become contaminated during the process.

- Reduce risks by removing highly hazardous food items from your menu. For example, you may want to avoid egg salad sandwiches if sandwiches must be transported and held before being served.

- Evaluate storage, prep, cooking and serving procedures to isolate any points where contamination might occur.

- Rank these hazards in terms of severity (how serious are the consequences) and probability (how likely are they to occur).

Step 2: Identify "Critical Control Points"

Identify the points in the process where hazards can be controlled or prevented. Develop a flowchart or create a step-by-step list detailing the preparation of each potentially hazardous food. Then, identify procedures to prevent, reduce and eliminate recontamination hazards at each step you have listed.

In general, food service workers can reduce the risk of food-borne illness by:

1. Practicing good personal hygiene.

2. Avoiding cross-contamination.

3. Using proper storage, cooking and cooling procedures.

4. Reducing the number of steps involved in preparing and serving.

Don't forget to review your vendor's food handling procedures. It is also your responsibility to see that food is properly transported and handled during delivery.

Step 3: Establish "Critical Limits"

To insure a food passes safely through a critical control point, you need to establish critical limits that must be met. These critical limits should be standards that are observable and measurable. They should include precise time, temperature and sensory requirements.

- Specify exactly what should be done to meet each particular standard. For example, instead of saying that a "food must be thoroughly cooked," the standard might say "heat rapidly to an internal temperature of 165° F within two hours."

- Make sure employees have calibrated, metal-¬stemmed or digital thermometers and that they use them routinely.

- Make sure recipes state the end cooking, reheating, and hot-holding temperatures; specific times for thawing, cooking and cooling foods.

- Schedule sufficient staff in peak hours to prepare and serve foods safely.

Step 4: Monitor the "Critical Control Points"

Using your flowcharts or lists, follow potentially hazardous foods through every step in the process. Compare your operation's performance with the requirements you have set. Identify any areas of deficiency.

Step 5: Take Corrective Action

Take corrective action as needed. For example, if products' temperatures are unacceptable when received, reject the shipment. Similarly, if…

- Food is contaminated by hands or equipment, rewash or discard it.

- Food temperature is not high enough after cooking, continue cooking to the required temperature.

- Food temperature exceeds 55 degrees F during cold prep or serving, discard it.

Step 6: Develop a Recordkeeping System

Develop a recordkeeping system to document the HACCP process and monitor your results. This may be any simple, quick system, such as a log, in which employees can record their compliance with standards at critical control points. These records are crucial and may provide proof that a food-borne illness did not originate at your establishment.

Step 7: Verify Your System's Effectiveness

Verify that the HACCP process in your facility works. You can do this in a number of ways.

- Be alert to how often you need to take corrective actions. If you need to take corrective actions frequently, this may indicate a need to change, or at least fine-tune, your system or retrain employees.

- Think of tests you can do like measuring the strength of your sanitizing solution. Also, examine your records and make sure employees are entering actual valid data.

- Make certain that your dishwashing and sanitizing equipment are working properly. Regular calibration and maintenance are a must.

- Ask your Board of Health for a non-biased assessment of whether or not your process is working.

HACCP Procedures

Purchasing

The goal of purchasing is to obtain wholesome, safe ingredients and prepared foods to meet your production and menu requirements. Safety at this step is primarily the responsibility of your vendors; however, it is your job to choose your vendors wisely.

Suppliers must meet federal and state health standards. They should use the HACCP system in their operations and train their employees in sanitation. Delivery trucks should have adequate refrigeration and freezer units. Foods should be packaged in protective, leak proof, durable packaging.

Let vendors know up front what you expect from them. Put food-safety standards in your purchase specification agreements. Ask to see their most recent Board of

Health Sanitation Reports and tell them you will be inspecting trucks on a quarterly basis.

Receiving

The goals of receiving are:

- To make sure foods are fresh and safe when they enter your facility.

- To transfer them to proper storage as quickly as possible.

Let's look more closely at two important parts of receiving:

- Getting ready to receive food.

- Inspecting the food when the delivery truck arrives.

There are several important guidelines to keep in mind and tasks to complete as you get ready to receive food:

- Make sure your receiving area is equipped with sanitary carts for transporting goods.

- Plan ahead for deliveries to ensure sufficient refrigerator and freezer space.

- Mark all items for storage with the date of arrival or the "use by" date.

- Keep the receiving area well lit and clean to discourage pests.

- Remove empty containers and packing materials immediately to a separate trash area.

- Keep all flooring clean of food particles and debris.

When the delivery truck arrives, make sure it looks and smells clean and is equipped with the proper food storage equipment. Then inspect foods immediately:

- Check expiration dates of milk, eggs and other perishable goods.

- Make sure shelf life dates have not expired.

- Make sure frozen foods are in airtight, moisture¬-proof wrappings.

- Reject foods that have been thawed and refrozen. Look for signs of thawing and refreezing such as large crystals, solid areas of ice or excessive ice in containers.

- Reject cans that have any of the following: swollen sides or ends; flawed seals or seams; dents or rust. Also, reject any cans whose contents are foamy or bad smelling.

- Check temperature of refrigerated and frozen foods, especially eggs and dairy products, fresh meat, and fish and poultry products.

- Look for content damage and insect infestations.

- Reject goods and foods delivered in flats or crates that are dirty.

Storing

In general, there are four possible ways to store food:

1. In dry storage, for longer holding of less perishable items.

2. In refrigeration, for short-term storage of perishable items.

3. In specially designed deep-chilling units for short periods.

4. In a freezer, for longer-term storage of perishable foods.

Each type of storage has its own sanitation and safety requirements.

Dry Storage

There are many items that can be safely held in a sanitary storeroom. These include, for example: canned goods, baking supplies (such as salt and sugar), grain products (such as rice and cereals) and other dry items. In addition, some fruits (such as bananas, avocados, and pears) ripen best at room temperature. Some vegetables, such as onions, potatoes and tomatoes, also store best in dry storage. A dry-storage room should be clean and orderly. The storage should be properly ventilated, kept at an appropriate temperature and humidity to retard the growth of bacteria and mold. Keep in mind the following:

- For maximum shelf life, dry foods should be held at 50 degrees F, but 60–70 degreesF is adequate for most products.

- Use a wall thermometer to check the temperature of your dry-storage facility regularly.

- To ensure freshness, store opened items in tightly covered containers. Use the "First In, First Out" (FIFO) rotation method, dating packages and placing incoming food items in the back so that older ones will be used first.

- To avoid pest infestation and cross-contamination, clean up all spills immediately and do not store trash or garbage cans in food storage areas.

- Do not place any items – including paper products – on the floor. Make sure the bottom shelf of the dry-storage room is at least 6 inches above the ground.

- To avoid chemical contamination, *never use or store cleaning materials or other chemicals where they might contaminate foods!* Store them, labeled, in their own section in the storeroom away from all food supplies. Always store chemicals in their original containers and post usage guidelines and warnings.

Refrigerated Storage

Keep fresh meat, poultry, seafood, dairy products, fruit, vegetables and hot leftovers in the refrigerator (below 40 degrees F).

Although no food can last forever, refrigeration increases the shelf life of most products. Most impor¬tantly, because refrigeration slows bacterial growth, the colder a food is, the safer it is.

Your refrigeration unit should contain open, slotted shelving to allow cold air to circulate around food. Do not line shelves with foil or paper. Also, do not over¬load the refrigerator and be sure to leave space between items to further improve air circulation.

All refrigerated foods should be dated and properly sealed. In addition:

- Use clean, nonabsorbent, covered containers that are approved for food storage.

- Store dairy products separately from foods with strong odors like onions, cabbage and seafood.

- To avoid cross-contamination, store raw or uncooked food away from and below prepared or ready-to-eat food.

- Never allow fluids from raw poultry, fish or meat to come into contact with other foods.

- Keeping perishable items at the proper temperature is a key factor in preventing food-borne illness. Check the temperature of your refrigeration unit regularly to make sure it stays below 40° F. Keep in mind that opening and closing the refrigerator door too often can affect temperature.

Many commercial refrigerators are equipped with externally mounted or built-in thermometers. These are convenient when they work, but it is important to have a backup.

It's a good idea to have several thermometers in different parts of the refrigerator to ensure consistent temperature and accuracy of instruments. Record the temperature of each refrigerator on a chart prefer¬ably once a day.

Deep Chilling

Deep or super chilling—that is, storing foods at temperatures between 26° F and 32° F—has been found to decrease bacterial growth. This method can be used to increase the shelf life of fresh foods, such as poultry, meat, seafood, and other protein items, without compromising their quality by freezing. You can deep-chill foods in specially designed units or in a refrigerator set to deep-chilling temperature.

Frozen Storage

Frozen meats, poultry, seafood, fruits and vegetables and some dairy products, such as ice cream, should be stored in a freezer at 0° F to keep them fresh and safe for an extended period of time.

As a rule, you should use your freezer primarily to store foods that are frozen when you receive them. Freezing refrigerated foods can damage the quality of perishable items. It's important to store frozen foods immediately. It's also important to remember that storing foods in the freezer for too long increases the likelihood of conta¬mination and spoilage. Like your refrigeration unit, the freezer should allow cold air to circulate around foods easily.

Be sure to:

- Store frozen foods in moisture-proof material or containers to minimize loss of flavor, as well as discoloration, dehydration and odor absorption.

- Monitor temperature regularly, using several thermometers to ensure accuracy and consistent temperatures. Record the temperature of each freezer on a chart.

- Open freezer doors only when necessary and remove as many items at one time as possible. You can also use a freezer "cold curtain" to help guard against heat gain.

- To minimize heat gain, place lower the temperature of warm foods in the refrigerator before storing in the freezer.

- Use the first-in-first-out (FIFO) inventory method to keep your inventory fresh.

- Date (with a freezer marker) "occasional use" items and toss anything beyond the maximum storage periods. Your freezer manufacturer or food vendor can supply you with their recommendations.

Preparing Food

Prepping

Most fruits and vegetables are purchased in a "straight from the field" condition. All should be properly washed to remove dirt, sand, twigs and hitchhiking insects. Soap and water is often sufficient to remove these and residual pesticides. You might also consider a food-safe disinfectant solution (as approved by your local Health Department) as an extra precaution for "high-risk" (elderly, children and immune impaired) customers. In preparing raw ingredients,

- Employees should sanitize their hands and work surfaces with food-safe chemicals before handling.

- All cutting instruments (knives, choppers, peelers) should be thoroughly sanitized between uses to eliminate cross-contamination.

- Produce with peels should be scrubbed before slicing or peeling to avoid transferring germs and chemicals to the edible portions.

Thawing and Marinating

Freezing food keeps most bacteria from multiplying, but it does not kill them. Bacteria that are present when food is removed from the freezer will multiply rapidly if thawed at an improper temperature.

Thus, it is critical to thaw foods *below* the "temperature danger zone." NEVER thaw foods on a counter or in any other unrefrigerated area!

Some foods, such as frozen vegetables, can be cooked from the frozen state. It is important to note, however, that this method depends on the size of the item. For example, cooking from frozen is not recommended for large food items.

The two best methods for thawing foods are:

1. In refrigeration at a temperature below 40 degrees F, placed in a pan on the lowest shelf so juices cannot drip on other foods.

2. Under clean, drinkable running water at a temperature of 70 degrees F or less for no more than two hours or just until the product is thawed.

ALWAYS marinate meat, fish and poultry in the refrigerator—NEVER at room temperature. NEVER save and reuse marinade. With all methods, be careful not to cross-contaminate.

Cautions for Cold Foods

When you are preparing cold ingredients, you are at one of the most hazardous points in the food-preparation process. There are two key reasons for this risk:

1. Cold food preparation usually takes place at room temperature.

2. Cold food is one of the most common points of contamination and cross-contamination.

Chicken, eggs, ground meat, produce salad, tuna salad, potato salad with eggs and other protein-rich salads are common sources of food-borne illness. Sandwiches prepared in advance and held unrefrigerated are also dangerous.

Because cold foods frequently require no further cooking, it is essential that all ingredients used in them are properly cleaned, prepared and, where applicable, cooked. It is a good idea to chill meats and other ingredients and combine them while chilled.

Here are several other important precautions to keep in mind:

- Prepare foods no further in advance than necessary.

- Prepare foods in small batches and place in cold storage immediately. This will prevent holding food too long in the "temperature danger zone."

- Always hold prepared cold foods below 40 degrees F.

- Thoroughly wash all fresh fruits and vegetables with plain water to remove surface pesticide residues and other impurities, such as soil particles.

- Don't overlook thick -produce such as avocadoes and cantaloupe. Once your knife penetrates the skin, you can introduce contaminates from the grower, from the work surface and knife and your hands.

- Use a brush to scrub thick-skinned produce, if desired.

Beware of **CROSS-CONTAMINATION**. It's crucial to:

- Sanitize your hands before and after handling each food item.

- Keep raw products separate from ready-to-serve foods.

- Sanitize cutting boards, knives and other food¬ contact surfaces after each contact with a potentially hazardous food.

- Use separate color-coded prep equipment to identify those used for produce, chicken and other at-risk items.

- Discard any leftover batter, breading or marinade after it has been used with potentially hazardous foods.

Cooking

Even when potentially hazardous foods are properly thawed, bacteria and other contaminants may still be present. Cooking foods to the proper internal temperature will kill any existing bacteria and make food safe.

It's important to remember, however, that conventional cooking procedures cannot destroy bacterial spores nor deactivate their toxins.

Keep in mind the following "safe cooking" tips:

- Stir foods cooked in deep pots frequently to ensure thorough cooking.

- When deep-frying potentially hazardous foods, make sure fryers are not overloaded, and make sure the oil temperature returns to the required level before adding the next batch. Use a hot-oil thermometer designed for this special application.

- Regulate size and thickness of each portion to make cooking time predictable and uniform.

- Allow cooking equipment to heat up between batches.

- Never interrupt the cooking process. Partially cooking poultry or meat, for example, may produce conditions that encourage bacterial growth.

- Monitor the accuracy of heating equipment with each use by using thermometers.

- Always use a thermometer to ensure food reaches the proper temperature during cooking. Use a sanitized metal-stemmed, numerically scaled thermometer (accurate to plus or minus 2 degrees F) or a digital thermometer.

- Check food temperature in several places, especially in the thickest parts, to make sure the food is thoroughly cooked. To avoid getting a false reading, be careful not to touch the pan or bone with the thermometer.

- Always cook food to an internal temperature of 165 degrees.

Serving and Holding

Food that has been cooked isn't necessarily safe. In fact, many outbreaks occur because improper procedures were used following cooking. Although it may be tempting to hold food at temperatures just hot enough to serve, it is essential to keep prepared foods out of the "temperature danger zone." This means, specifically:

- Always keep HOT foods in hot-holding equipment above 140 degrees F.

- Always keep COLD foods in a refrigeration unit or surrounded by ice below 40 degrees F.

For safer serving and holding:

- Use hot-holding equipment, such as steam tables and hot-food carts, during service but never for reheating.

- Stir foods at reasonable intervals to ensure even heating.

- Check temperatures with a food thermometer every 30 minutes.

- Sanitize the thermometer before each use or use a digital infrared thermometer that never touches the food.

- Cover hot-holding equipment to retain heat and to guard against contamination.

- Monitor the temperature of hot-holding equipment with each use.

- DISCARD any food held in the "temperature danger zone" for more than 4 hours!

- To avoid contamination, *never add fresh food to a serving pan containing foods* that have already been out for serving!

Some key points...

1. Always wash hands with soap and warm water for at least 20 seconds before serving food.

2. Use cleaned and sanitized long-handled ladles and spoons so bare hands do not touch food.

3. Never touch the parts of glasses, cups, plates or tableware that will come into contact with food.

4. Never touch the parts of dishes that will come into contact with the customer's mouth.

5. Wear gloves if serving food by hand.

6. Cover cuts or infections with bandages, and if on hands, cover with gloves.

7. Discard gloves whenever they touch an unsanitary surface.

8. Use tongs or wear gloves to dispense rolls and bread.

9. Clean and sanitize equipment and utensils thoroughly after each use.

10. Use lids and sneeze guards to protect prepared food from contamination.

11. Keep cash handling duties separate from food handling whenever possible.

12. **To avoid contamination** – always wash hands, utensils and other food-contact surfaces *after* contact with raw meat or poultry and *before* contact with cooked meat or poultry.
    o For example, do not reuse a serving pan used to hold raw chicken to serve the same chicken after it's cooked, unless the pan has been thoroughly cleaned and sanitized.

Sanitary Self-Service

Like workers, customers can also act as a source of contamination. Unlike workers, customers - especially children - are, generally, not educated about food sanitation and may do the following unsanitary things:

- Touch food and open plastic ware with their hands.

- Touch the edges of eating/serving utensils and equipment such as soda dispensers.

- Sneeze or cough into food or self-serve displays.

- Touch salt-and-pepper shaker tops, sugar bowls and condiment containers.

- Return food items to avoid waste.

- Observe customer behavior and remove any foods that may have been contaminated.

- Serve sealed packages of crackers, breadsticks and condiments.

- Prewrap, date and label sandwiches if possible.

Cooling

Here, as at other critical points, every move you make can mean the difference between the safe and the unsafe.

It is often necessary to prepare foods in advance or use leftover foods. Unfortunately, this can easily lead to problems, unless proper precautions are taken. In fact, problems at this stage are the number-one cause of food-borne illness. The two key precautions for preventing food-borne illness at this point in the process are rapid cooling and protection from contamination.

Chilling It Quickly

All potentially hazardous, cooked leftovers should be chilled to an internal temperature of below 40° F. Quick-chill any leftovers larger than half a gallon or 2 pounds.

Quick chilling involves five simple steps:

- **Reduce food mass.** Smaller amounts of food will chill more quickly than larger amounts, so cut large items into pieces or divide food among several containers or shallow pans.

- **Use shallow, prechilled pans** (no more than 4 inches deep). Use stainless-steel containers when possible; stainless steel transfers heat better and cools faster than plastic.

- **Chill.** Ideally, place food in an ice-water bath or quick-chill unit (26–32° F) rather than a refrigerator. These options are best for two reasons:
  - o   Water is a much better heat conductor than air. As a result, foods can cool much more quickly in an ice bath than they can in a refrigerator.
  - o   Refrigeration units are designed to keep cold foods cold rather than to chill hot foods. They can take too long to cool foods to safe temperatures.

- **Prechill foods in a freezer** for about 30 minutes before refrigerating.

- **Separate food items** so air can flow freely around them. Do not stack shallow pans. NEVER cool at room temperature.

- **Stir frequently.** Stirring accelerates cooling and helps to ensure that cold air reaches all parts of the food.

- **Measure temperature periodically.** Food should reach a temperature of 70 degrees F within two hours and 40 degrees F within four hours. It's important to note that this time must be reduced if food has already spent time in the "temperature danger zone" at any other point in the preparation and serving process.

- **Tightly cover and label cooled foods.** On labels, include preparation dates and times.

- **Store** uncovered cooked and cooled foods on the upper shelves of the cooler and cover them when they reach 45degrees F.
  - o   Be aware that although uncovered foods cool faster, they are at increased risk for cross-contamination.
  - o   Never store them beneath raw foods.

Reheating

While assuming leftovers are safe might seem reasonable, it's not. In reheating and serving leftovers—just as in all phases of the food-preparation process—you must be careful to avoid contamination.

To safely reheat and serve leftovers, be sure to:

- Boil sauces, soups and gravies and heat other foods to a minimum of 165° F, within two hours of taking the food out of the refrigerator.
- Never reheat food in hot-holding equipment.
- Never mix a leftover batch of food with a fresh batch of food.
- Never reheat food more than once.

The Difference Between Clean And Sanitary

Heat or chemicals can be used to reduce the number of bacteria to acceptable levels. They can also be used combat other harmful microorganisms.

Heat sanitizing involves exposing equipment to high heat for an adequate length of time. This may be done manually by immersing equipment in water maintained at a temperature of 170–195° F for at least 30 seconds or in a dishwashing machine that washes at 150° F and rinses at 180° F.

For any method, it is important to check water tem¬perature frequently. Thermometers and heat-sensi¬tive tapes and labels are available for determining whether adequate sanitation temperatures have been achieved.

Immerse an object in or wipe it down with bleach or sanitizing solution. For bleaching purposes, use ½ ounce or 1 tablespoon of 5-percent bleach per gallon of water. For using commercial products, follow the manufacturers' instructions.

Chemical sanitizers are regulated by the EPA. Manufacturers must follow strict labeling require-ments regarding what concentrations to use, data on minimum effectiveness and warnings of possible health hazards.

Chemical test strips are available for testing the strength of the sanitizing solution. Because sanitiz¬ing agents become less effective as they kill bacteria and are exposed to air, it is important to test the sanitizing solution frequently.

Sanitizing Portable Equipment:

- To properly clean and sanitize portable equipment you must have a sink with three separate compartments: for cleaning, rinsing and sanitizing.

- There should be a separate area for scraping and rinsing food and debris into a garbage container or disposer before washing, and separate drain boards for clean and soiled items.

- To sanitize a piece of equipment, use the following procedure:

  1. Clean and sanitize sinks and work surfaces.

  2. Scrape and rinse food into garbage or disposal. Presoak items, such as silverware, as necessary.

  3. In the first sink, immerse the equipment in a clean detergent solution at about 120 degrees F. Use a brush or a cloth to loosen and remove any remaining visible soil.

  4. Rinse in the second sink using clear, clean water between 120° F and 140° F to remove all traces of food, debris and detergent.

  5. Sanitize in the third sink by immersing items in hot water at 170° F for 30 seconds or in a chemical sanitizing solution for one minute. Be sure to cover all surfaces of the equipment with hot water or the sanitizing solution and keep them in contact with it for the appropriate amount of time.

  6. If soapsuds disappear in the first compartment or remain in the second, if the water temperature cools, or if water in any compartment becomes dirty and cloudy, empty the compartment and refill it.

  7. Air dry. Wiping can recontaminate equipment and can remove the sanitizing solution from the surfaces before it has finished working.

  8. Make certain all equipment is dry before putting it into storage; moisture can foster bacterial growth.

Sanitizing In-Place Equipment:

Larger and immobile equipment should also be washed, rinsed and sanitized.

Use the following procedure:

1. Unplug electrically powered equipment, such as mixers.

2. Remove fallen food particles and scraps.

3. Wash, rinse and sanitize any removable parts using the manual immersion method described in steps 3 through 5 in the prior list.

4. Wash the remaining food-contact surfaces and rinse with clean water.
    Wipe down with a chemical sanitizing solution mixed according to the
    manufacturer's directions.

5. Wipe down all non-food contact surfaces with a sanitized cloth, and allow
    all parts to air dry before reassembling. Sanitize cloth before and during
    sanitizing by rinsing it in sanitizing solution.

6. Resanitize the external food-contact surfaces of the parts that were handled
    during reassembling.

7. Scrub wooden surfaces, such as cutting boards, with a detergent solution
    and a stiff-bristled nylon brush, then rinse in clear, clean water and wipe
    down with a sanitizing solution after every use.

A First-Rate Facility

Safe and sanitary food service begins with a facility that is clean and in good repair.
The entire facility – work areas as well as equipment – should be designed for easy
cleaning and maintenance.

It's important to eliminate hard-to-clean work areas, as well as faulty or overloaded
refrigerators or other equipment. Also, get rid of dirty surroundings and any conditions that will attract pests.

Remember – the easier the workplace is to clean, the more likely it will stay clean.

Work Areas

Work areas should be made from non-porous materials that can withstand chemical
and/or steam cleaning. Look for well-constructed tables and counters without joints
or seams that can be hard to clean. Flat worktables with locking wheels can be useful
during peak work periods.

Hand sinks

Hand sinks should be conveniently located near work areas. To avoid cross-contamination, consider faucets and soap dispensers that are hands-free (elbow or foot control). Use paper towels, as these are the more sanitary than other hand drying methods.

Floors, Walls and Ceilings

• Floors, walls and ceilings should be free of dirt, litter and moisture.

- Clean walls regularly by swabbing with a cleaning solution or by spraying with a pressure nozzle. Sweep floors, then clean them using a spray method or by mopping.

- Swab ceilings, instead of spraying them, to avoid soaking lights and ceiling fans.

- Don't forget corners and hard-to-reach places!

Ventilation

Good ventilation is a critical factor in maintaining a clean food-service environment. Ventilation removes steam, smoke, grease and heat from food-preparation areas and equipment. This helps maintain indoor air quality and reduces the possibility of fires from accumulated grease. In addition, good ventilation eliminates condensation and other airborne contaminants. It also:

- Reduces the accumulation of dirt in the food-preparation area.

- Reduces odors, gases and fumes.

- Reduces mold growth by reducing humidity.

To ensure good ventilation, be sure to:

- Use exhaust fans to remove odors and smoke.

- Use hoods over cooking areas and dishwashing equipment.

- Check exhaust fans and hoods regularly to make sure they are clean and operating properly.

- Clean hood filters routinely according to the instructions provided by the hood manufacturer.

Storerooms

Like all areas of the facility, storerooms must be kept clean and litter-free. To accomplish this, be sure to sweep and scrub walls, ceilings, floors, shelves, light fixtures and racks on a routine basis. Check all storage areas frequently—this includes your refrigerator and freezer as well as your dry-storage room. In checking storage areas:

1. Look for damaged or spoiled foods, broken or torn packages and bulging or leaking cans.

2. Remove any potentially spoiled foods immediately, and clean the area thoroughly.

3. Make sure foods and other supplies are stored at least 6 inches from the walls and above the floor.

4. Store cleaning supplies and chemicals in a separate area away from food supply areas and other chemicals so they do not pose a hazard to food or people.

Restrooms

Restrooms public and private should be convenient, sanitary and adequately stocked with the following:

- Toilet paper

- Antiseptic liquid soap

- Disposable paper towels and/or air blowers

- Covered trash receptacles (The trash receptacle lid should open with a foot pedal.)

Check public restrooms throughout the day and scrub daily.

In private employee restrooms, you may also want to provide brushes to wash finger-nails and sanitizing solution for soaking the brushes.

Avoid Cross-Contamination

One of the most common causes of food-borne illness is cross-contamination: the transfer of bacteria and viruses from food to food, hand to food or equipment to food.

Food to Food: Raw, contaminated ingredients may be added to foods, or fluids from raw foods may drip into foods that receive no further cooking. A common mistake is to leave thawing meat on a top shelf in the refrigerator where it can drip down onto prepared foods stored below.

Hand to Food: Bacteria are found throughout the body: in the hair, on the skin, in clothing, in the mouth, nose and throat, in the intestinal tract and on scabs or scars from skin wounds. These bacteria often end up on the hands where they can easily spread to food. People can also pick up bacteria by touching raw food, then transfer it to cooked or ready-to-eat food.

Equipment to Food: Bacteria may pass from equipment to food when equipment that has touched contaminated food is then used to prepare other food without proper cleaning and sanitizing. For example, cross-contamination can occur when surfaces used for cutting raw poultry are then used to cut foods that will be eaten raw, such as fresh vegetables.

Coverings, such as plastic wrap and holding and serving containers, can also harbor bacteria that can spread to food. A can opener, a plastic-wrap box or a food slicer can also become a source of cross-contamination if not properly sanitized between uses.

Personal Hygiene

Personal hygiene is the best way to stop bacteria from contaminating and spreading into new areas. Hands are the greatest source of contamination. Hands must be washed constantly throughout the day. Every time an individual scratches her head or sneezes, she is exposing her hands to bacteria and will spread it to anything she touches, such as food, equipment and clothes. Hand and nail brushes, anti-bacterial soaps and disposable gloves should be a part of every bakery, even if not required by law. Proper training and management follow-up is also critical.

Every employee must practice good basic hygiene:

    A.   Short hair, and/or hair contained in a net.

    B.   Clean shaven, or facial hair contained in a net.

    C.   Clean clothes/uniforms.

    D.   Clean hands and short nails.

    E.   No unnecessary jewelry.

    F.   A daily shower or bath.

    G.   No smoking in or near the kitchen.

An employee who has the symptoms of the common cold or any open cuts or infections should not go to work. By simply breathing, he may be inadvertently exposing the environment to bacteria. Although it is rarely practiced in the food industry, all employees should be required to have a complete medical examination as a condition of employment. This should include blood and urine tests. A seemingly healthy individual may unknowingly be the carrier of a latent communicable disease.

Are Your Hands Really Clean?

Hand washing is perhaps the most critical aspect of good personal hygiene in food service. Workers should wash their hands with soap and warm water for 20 seconds. (Hint: Sing "Happy Birthday" twice.)

When working with food, they should wash gloved hands as often as bare hands. Hand washing is such a simple—yet very effective—method for eliminating cross-contamination. Employees should be trained to wash their hands periodically throughout the day. Hand washing must be done:

1. Before starting work

2. After handling any foreign object: head, face, ears, money, food, boxes or trash.

The following exercise will reinforce your hand washing procedures and makes a great interactive training activity during a staff meeting.

First, you'll need a fluorescent substance and a black light. A convenient kit is Atlantic Publishing's Glo Germ Training Kit available at **www.atlantic-pub.com** or call 800-541-1336. Using these materials, you can show trainees the "invisible dirt" that may be hiding on their hands:

A. Have employees dip their hands in the fluorescent substance.

B. Tell employees to wash their hands.

C. Have employees hold their hands under the black light to see how much "dirt" is still there.

D. Explain proper hand-washing technique.

E. Have employees wash their hands again, this time using the proper hand-washing technique.

F. Have employees once again hold their hands under the black light.

The Intruders

Your safety precautions and proper food handling are all done to eliminate or neutralize potential contaminants spread by humans and animals. Below you'll find some basic information on food-borne bacteria, human contact viruses and pests.

Bacteria

Bacteria are everywhere: in the air, in all areas of the bakery and all over one's body. Most bacteria are microscopic and of no harm to people. Many forms of bacteria are actually beneficial, aiding in the production of such things as cheese and sourdough bread. Only a small percentage of bacteria will cause food to spoil and can generate a form of food poisoning when consumed.

Bacteria need food, water and warmth in order to survive. Bacteria prefer to ingest moisture-saturated foods, such as meats, dairy products and produce. They will not grow as readily on dry foods such as cereals, sugar or flour. Bacteria will grow most rapidly when the temperature is between 85-100 degrees F. In most cases, the growth rate will slow down drastically if the temperature is hotter or colder than this. Thus, it is vitally important that perishable food items are refrigerated before bacteria have a chance to establish themselves and multiply. Certain bacteria can survive in extreme hot- and cold-temperature ranges. By placing these bacteria in severe temperatures, you will be slowing down their growth rate, but not necessarily killing them.

Bacterial forms do not have a means of transportation; they must be introduced to an area by some other vehicle. People are primarily responsible for transporting bacteria to new areas. The body temperature of 98.6 degrees F is perfect for bacterial existence and proliferation. A person coughing, sneezing or wiping their hands on a counter can introduce bacteria to an area. Bacteria may be transmitted also by insects, air, water and articles onto which they have attached themselves, such as boxes, blades, knives and cutting boards.

Controlling Bacteria

The first step in controlling bacteria is to limit their access to the bakery. Make certain that all products entering the bakery are clean. Follow the prescribed bug-exterminating procedures to stop bacteria from being transported into the bakery. Keep all food products stored and refrigerated as prescribed. Clean up any spills as you go along, making the environment unsuitable for bacteria to live. Keep all food refrigerated until needed, and cook it as soon as possible.

The quality known as "pH" indicates how acidic or alkaline ("basic") a food or other substance is. The pH scale ranges from 0.0 to 14.0—7.0 being exactly neutral. Distilled water, for example, has a neutral pH of 7.0. Bacteria grow best in foods that are neutral or slightly acidic, in the pH range of 4.6 to 7.0. Highly acidic foods, such as vinegar and most fresh fruits, inhibit bacterial growth. Meats and many other foods have an optimal pH for bacterial growth. On the other hand, some foods normally considered hazardous, such as mayonnaise and custard filling, can be safely stored at room temperature if their pH is below 4.6.

Handling Viruses

Food-borne Hepatitis A is an infectious virus that affects the liver. The virus enters through the mouth, multiples within the body and passes to the feces. The virus is commonly transmitted via the hands because of poor personal hygiene and diapering habits. Flies can also spread the virus by landing on food. Drinking water and shellfish can also be carriers of this food-borne virus. Because the time between exposure and symptoms can be as long as 50 days, carriers have plenty of time to infect others.

Children as especially susceptible; although they may be without common symptoms.

The Hepatitis A virus can be inactivated by boiling or cooking to 185 degrees F. However, contamination can occur after cooking. Proper hygiene and cleaning of foods is your best defense against Hepatitis A. Hands must be thoroughly washed after using the restroom. However, because the virus can remain active for up to 30 days on surfaces. Door knobs, faucets and other surfaces can contaminate sanitized hands. It is critical that hands be sanitized AGAIN before and after handling food. Utensils and work surfaces must also be sanitized regularly. Infected employees must be reported to your Health Department and placed on a leave of absence until released by a physician.

Bugs, Insects And Animal Pests

Bug, insect and rodent infestation in a bakery is the result of poor sanitation practices. Aside from being a nuisance, they are a threat to food safety. Flies, cockroaches and other insects all carry bacteria, and many, because of where they get their food, carry disease.

Pests require the same three basic necessities of life as bacteria do: food, water and warmth. When healthful, thriving bugs and insects are visible, this is an indicator that proper sanitation procedures have not been carried out. Eliminate the environment that these pests need to live, and you will be eliminating their existence. Combining proper sanitation practices with periodic food-safe extermination will stop any problems before they start.

Prevention

All doorjambs and building cracks, even the thinnest ones, must be sealed. Be cautious when receiving deliveries. Bugs may be in the boxes or crates.

To prevent the spread of flies in your establishment,

- Keep all doors, windows and screens closed at all times.

- Ensure that garbage is sealed in airtight containers and is picked up regularly. All trash must be cleaned off the ground: flies can deposit their eggs on the thinnest scrap of food.

- Dumpsters must be periodically steam cleaned and deodorized. They should never contain any decaying food scraps.

The greatest protection against cockroaches is your exterminator. Of course, the exterminator will be of little value, if you do not already have good sanitary practices in place. Select an exterminator who is currently servicing other bakeries. Chemicals sprayed in a bakery must be of the nonresidual type. These are safe and approved for use in food-service establishments.

Animal pests, such as rats and mice, may be very serious problems for the bakery operator. Rodents are prolific breeders, producing as many as 50 offspring in a lifespan of one year. They tend to hide during the day but can be discovered by telltale signs.

Rats and mice are filthy animals that will eat any sort of garbage or decaying food available; and are disease and bacteria carriers. They have been known to bite people, as have their fleas, which also spread their bacteria and disease.

They are extremely strong and can easily gain access to a building through a crack or hole no larger than a quarter. Ensure that your building's foundation is airtight. Keep all food products at least 6 inches off the floor; this enables the exterminator to get under the shelving to spray. Rat bait, a poisoning capsule resembling food, is particularly effective when spread around the building and dumpsters. As with any poison or chemical you use, make certain that it is labeled clearly and stored away from food-storage areas.

Kitchen Safety

By its nature, the food service environment is full of potential hazards to employees' safety. Knives, slicers, grinders, glass, hot surfaces and wet or greasy floors are only a few of the hazards food-service workers face every day. Fortunately, most accidents also involve human error and, therefore, can be prevented.

Careful—It's Hot

There are many ways employees can be burned in a food-service environment unless they're very careful. Burns can result from contact with hot surfaces such as grills, ovens, burners, fryers and other heating equipment. Burns can also be caused by escaping steam or by hot food or drinks that are splattered, splashed or spilled.

To prevent burns...

- Use thick, dry potholders or mitts, and stir food with long-handled spoons or paddles.

- Turn on hot-water faucets cautiously. Wear insulated rubber gloves for rinse water that is 170 degrees F. Follow instructions for the use of cooking equipment—particularly steam equipment. Be sure all steam is expelled from steamers before opening the doors.

- Lift cooking lids and similar equipment away from yourself to avoid burns from steam.

- To avoid splattering and splashing, don't fill kettles too full. Also, don't allow food to boil over.

- Remember that oil and water don't mix, so be sure food is DRY before you place it in a fryer.

- Point panhandles away from foot traffic, but also within reach, to avoid knocking over other pans.

- Do not crowd cooking surfaces with hot pans. Remove cooked foods from cooking surfaces immediately.

- Allow oil to cool and use extreme caution when cleaning fryers.

- Use caution when removing hot pans from the oven. Wear insulated gloves or mitts, and be certain no one is in the removal path.

- Do not wear clothing that may drape onto a hot spot and catch on fire.

Cuts and Abrasions

Just as they need to take precautions to prevent being burned, food-service workers also need to be careful not to get cut. And it's not just knives that can cause trouble.

Workers can hurt themselves – or their co-workers – with the sharp edges of equipment and supplies or with broken glass. Nails and staples used in food packaging can also be dangerous.

To prevent cuts, take the following precautions...

- Use appropriate tools (not bare hands) to pick up and dispose of broken glass. Immediately place broken glass into a separate, clearly marked garbage container.

- Take care when cutting rolls of kitchen wrap with the cutter.

- Be careful with can openers and the edges of open cans. Never use a knife to open cans or to pry items loose.

- Have box cutters available in receiving, storage and prep areas.

- Use a pusher to feed food into a grinder.

- Turn off and unplug slicers and grinders when removing food and cleaning.

- Use guards on grinders and slicers.

- Replace equipment blades as soon as they are cleaned.

- Be aware that left-handed people need to take extra care when working with slicers and similar equipment. This is because the safety features on this equipment are designed for right-handed people.

In addition...

- Keep knives sharp. Dull blades are harder to work with and cause more cuts than sharp ones.

- Never leave knives or equipment blades in the bottom of a sink.

- Carry knives by the handle with the tip pointed away from you. Never try to catch a falling knife.

- Cut away from yourself on a cutting board.

- Slice; do not hack.

Also, when you're storing or cleaning equipment, be sure to...

- Store knives and other sharp tools in special places when not in use.

- Wash dishes and glasses separately to help prevent them from being crushed by heavier objects and breaking in the dishwasher or sink.

- Do not stack glasses or cups inside one another.

- Watch out for nails, staples and protruding sharp edges while unpacking boxes and crates.

Electrical Shock

Because of the variety of electrical equipment used in food service, electrical shock is a common concern.

To prevent electrical shock...

- Properly ground all electrical equipment.

- Ensure that employees can reach switches without touching or leaning against metal tables or counters.

- Replace all worn or frayed electrical cords.

- Use electrical equipment only when hands are dry.

- Unplug equipment before cleaning.

- Locate electrical switches and breakers to permit rapid shutdown in the event of an emergency.

Strains

Carrying equipment or food items that are too heavy can result in strains to the arms, legs or back.

To prevent strains...

- Store heavy items on lower shelves.

- Use dollies or carts when moving objects that are too heavy to carry.

- To move objects from one area to another, use carts with firm shelves and properly operating wheels or casters.

- Don't carry too many objects at one time; instead, use a cart.

- Don't try to lift large or heavy objects by yourself.

- Use proper lifting techniques. Remember to bend from your knees, not your back.

Slipping And Falling

Anyone who slips and falls onto the floor can be badly hurt. Be sure your facility does not have hazards that put workers and customers at risk.

To prevent slips and falls...

- Clean up wet spots and spills immediately.

- Let people know when floors are wet. Use signs that signal caution, and prominently display them. Wear shoes that have no-slip soles.

- Do not stack boxes or other objects too high. They can fall and cause people to trip.

- Keep items such as boxes, ladders, step stools and carts out of the paths of foot traffic.

Fires

More fires occur in food service than in any other type of operation. Fire extinguishers should be available in all areas where fires are likely, especially in the kitchen near grills and deep fryers. But be careful—don't keep extinguishers so close to the equipment that they will be inaccessible in the event of a fire.

All employees should be trained in avoiding fires as well as in the use of fire extinguishers and in evacuation procedures. REMEMBER...Always call the fire department first, before using a fire extinguisher!

Choking

Anyone can choke on food if he or she is not careful. That's why an important part of food service safety is being alert to your customers.

Here's what to look for and what to do...

- If a person has both hands to the throat and cannot speak or cough, it is likely he or she is choking.

- If this person can talk, cough or breathe, do not pat him or her on the back or interfere in any way.

- If this person cannot talk, cough or breathe, you will need to take action. Use the Heimlich maneuver, and call for help immediately.

All food service employees should be trained in the use of the Heimlich maneuver. Posters with instructions on how to perform it should be posted near the employee dining area.

Exposure To Hazardous Chemicals

Improper exposure to cleaning agents, chemical pesticides and chemical sanitizers may cause injury to the skin or poisoning. To protect workers from exposure to hazardous materials, special precautions need to be taken, including certain steps that are required by law.

For example, the U.S. Department of Labor's Occupational Safety and Health Administration – commonly known as OSHA – requires food-service establishments to keep a current inventory of all hazardous materials.

Manufacturers are required to make sure hazardous chemicals are properly labeled and must supply a Material Safety Data Sheet (MSDS) to be kept on file at the food-service facility. The MSDS provides the chemical name of the product and physical hazards, health hazards and emergency procedures in case of exposure.

Information about each chemical – including its com¬mon name, when it is used, who is authorized to use it, and information from the MSDS – must also be pro¬vided to workers.

To prevent improper exposure to hazardous materials, make sure...

- only properly trained workers handle hazardous chemicals.

- employees have safety equipment to use when working with hazardous chemicals.

- employees wear nonporous gloves and eye goggles when working with sanitizing agents and other cleaners.

Safety First

Improper handling of food products or neglecting sanitation and safety procedures will certainly lead to health problems and/or personal injury. A successful bakery must develop a reputation for serving quality food in a safe environment. Should there ever be a question in your customers' minds as to the wholesomeness or quality of a product, the bakery will quickly lose its hard-earned reputation. The sanitation and safety procedures described in this section are very simple to initiate, but management must follow up and enforce them.

Chapter 18

Your Bakery Staff

· ·

Every bakery is unique in the way it operates. In a small independent bakery, people will probably handle a broad range of duties with each working for a common goal. Before you hire your first employee, it is important that you write a job description. Remember you cannot find the right person for the job if you don't really know what that job will entail.

Kitchen Personnel

Artisan bakeries focus primarily on the hands-on aspect of product creation while production bakeries focus on volume. Your bakery style will directly influence the artistic skills, production expertise and job responsibilities required of your baking staff.

Head Baker

In an artisan bakery, your head baker (you or someone else) will probably manage all aspects of production, act as the personnel director and be the quality control manager. In a production bakery, your head baker will probably be experienced in a variety of bakery support positions but their primary responsibility will be management of inventory, production processes and other personnel.

Bakery Manager

In larger bakery operations, or where food is served (a deli or café) or when the bakery is part of another retail operation, a Bakery Manager may be responsible for:

1. Hiring, training, supervising and scheduling employees.
2. Establishing and maintaining food quality control.
3. Controlling waste and food cost.
4. Ordering, receiving, storing and issuing all food products.
5. Health and safety regulation enforcement.
6. Setting safety standards, training employees and complying with local regulations.
7. Keeping owner informed of possible problem areas.
8. Acting as media contact.
9. Enhancing employee communications and morale.

Journey Baker

Larger bakeries may have journey bakers who assist directly in the production of baked goods. They may also function as a prep worker responsible for stocking ingredients, preparing equipment, measuring ingredients, mixing and baking products, finishing and clean-up. There are certified programs for journey bakers or you may develop this person in-house. The primary duties are the same as the Head Baker except for management responsibilities.

Short-order/Prep Cooks

If you have deli, snack bar or café service in your bakery, you'll may have prep cooks in the kitchen. These people are responsible for preparing simple entrees and side dishes for take-out or dine-in customers. Depending upon your menu offerings, you may require a dedicated staff member for these duties or other kitchen staff may handle these tasks.

Preparation Workers

Prep workers are support personnel for bakers, decorators and cooks. Their primary responsibility is to prepare all the ingredients in the bakery in accordance with the preparation methods prescribed. The head baker or kitchen manager trains, supervises and is responsible for the prep workers. In smaller operations, your prep workers may double as journey bakers.

The greatest amount of waste occurs during prep duties. The importance of conscientious staff cannot be overstated. By thoroughly training, monitoring and mentoring, you will be saving yourself potentially thousands of dollars annually and developing a promotable employee.

Your prep staff must follow the Recipe and Procedure Manual exactly as it is printed in order to ensure consistent products and control food costs.

Cake Decorators and Finishers

Cake decorators and finishes are your "detail" staff that prepare and finish sweet baked goods. Their duties combine technical experience and artistic flair. They are responsible for preparing icings, decorating a variety of cakes, creating displays and working with customers for personalized products.

Dishwasher

Your dishwashers may actually only wash a few dishes when you have paper only service. But washing your industrial size pots and pans along with utensils and tools can be a

Depending upon your staffing needs, journey bakers or prep workers may also double as your ware washer. While often considered the "lowliest" task in the kitchen, you have an opportunity as an employer to provide young workers with an opportunity to learn and grow. By building in responsibilities and listening to feedback, the person who scrubs your pots and keeps things clean and sanitary can be a valuable member of your team.

Dishwasher responsibilities could also include monitoring hot water/chemical output of equipment; tracking cleaning supply inventory; reporting breakage and damage of dishware and cookware; maintaining waste management and recycling program; and policing waste disposal areas.

Delivery Personnel

Wholesale operations and custom wedding cake bakers will need delivery personnel. Route drivers would typically handle wholesale deliveries during early morning hours. They would be required to load delivery van/trucks, delivery ordered products, interact with customers including handling customer complaints, provide customers with appropriate paperwork and oversee general vehicle maintenance. Minimum physical requirements would be established and training in proper lifting procedures would be important.

Wedding cake specialists frequently provide delivery and set-up at the reception hall or home. Elaborate cakes can require last-minute assembly and finishing on site. The delivery personnel should be able to handle minimum lifting requirements along with skilled in decorative assembly and presentation. Wedding and grooms' cakes can include special presentations that feature fresh flowers, unique packaging and serving advice. The delivery team would work closely with the bridal consultant and family.

Sales Manager

If you are interested in having wholesale accounts or providing custom products to consumers, a sales manager might be a wise choice. This person will work directly with wholesale food service customers interested in serving your breads and desserts in their restaurants, schools or corporate cafeterias or private labeling baked goods for retail sales. If a significant part of your business is creating custom products for consumers such as wedding cakes, your sales manager would be the person who meets with brides-to-be, caterers and wedding consultants. The sales manager can also provide your bakery with customer service management and training for your in-house staff.

Front Counter

Your retail operation will be judged by how your counter staff behaves. *Your business is only as good as the people who interact with your customers are.* While they may sweep the floor or restock the racks, their primary duty is to wait on customers in a courteous, friendly and helpful manner. Everything else becomes secondary when someone enters.

Your front counter staff will have a variety of duties related to keeping a "store" working smoothly. They must have strong cash handling and people skills. Your counter staff may also do final prep/cooking of ready-to-eat entrees and side dishes.

Chapter 19

Successful Kitchen Management and Control Procedures

• •

$\mathcal{M}$ anaging your kitchen requires good organizational skills, the capacity to analyze and project needs and, most importantly, the ability to supervise and motivate others.

Purchasing

The goal of purchasing is to supply the bakery with the best goods at the lowest possible cost. There are many ways to achieve this. You must have favorable working relations with all suppliers and vendors. A large amount of time must be spent meeting with prospective sales representatives and companies. Your responsibility is to evaluate and decide how to best make each of the purchases for the bakery. Purchasing is a complex area that must be managed by someone who is completely familiar with all of the bakery's needs. The kitchen director or bakery manager would be the best choice to do the purchasing.

It is preferable to have one person (and a trained backup) to do all the purchasing for all areas of the bakery. There are several advantages to this, such as greater negotiating power and better overall control.

Inventory Levels

The first step in computing what item and how much of it to order is to determine your minimum and maximum inventory levels. Your minimum level is your signal to reorder before you run out. Understanding your daily, weekly or monthly usage and delivery times will help you keep the proper quantity on hand. Your maximum inventory level also takes into account your usage data but it alerts your buyer from sinking too much into standing inventory. This process is often called "Just-In-Time" inventory/production. Tying up your valuable cash can negate any volume purchase discount you may receive. The key is to keep stocked without overstocking!

To determine the amount you need to order, you must first take a physical inventory on a regular basis. You may also rely on computerized inventory systems as a basis for purchasing; however, an actual count should be taken at least at the end of every quarter. Walk through the storage areas and mark in the "On Hand" column the amounts that are there.

To determine the "Desired Inventory," you will need to know when regularly scheduled deliveries arrive for that item and the amount used in the period between deliveries. Add on about 25 percent to the average amount used; this will cover unexpected usage, a late delivery or a backorder at the vendor.

The amount you need to order is the difference between the "Desired Inventory Level" and the amount "On Hand." Experience and food demand will reveal the amount an average order should contain.

A buying schedule should be set up and adhered to. This would consist of a calendar showing:

- Which day's orders need to be placed?
- When deliveries will be arriving.
- What items will be arriving from which company.
- Contact information for customer service and/or sales representatives to for each vendor.
- The price the sales representative quoted or other special pricing/discounts.

Post the buying schedule on the office wall. When a delivery doesn't arrive as scheduled, you should call the company immediately. Don't wait until the end of the day when offices are closed.

A **Want Sheet** may be placed on a clipboard in the kitchen. This sheet is made available for employees to write in any items they may need to do their jobs more efficiently. This is a very effective form of communication; employees should be encouraged to use it. You should consult this sheet every day.

A request might be as simple as a commercial-grade carrot peeler. If, for example, the last one broke, and the preparation staff has been using the back of a knife instead, the small investment could save you from an increase in labor and food costs.

Cooperative Purchasing

Many bakeries have formed cooperative purchasing groups to increase their purchasing power of bulk ingredients or specialty ingredients such as organic products delivered straight from the farm. By working together to place large orders, bakeries can usually get substantial price reductions. Some organizations even purchase their own trucks and warehouses and hire personnel to pick up deliveries. This can be quite advantageous for bakeries that are in the proximity of a major supplier or shipping center. Many items, such as produce, dairy products, seafood and meat, may be purchased this way. Franchise operations frequently have a centralized purchasing department and, often, large self-distribution centers that serve your region.

Receiving and Storing

Most deliveries will be arriving at the bakery during the day. Deliveries should only be received during set hours – typically after morning rush is complete and before lunch. The preparation crew is normally responsible for receiving and storing all items.. The buyer should also be present to ensure that each item is of the specification ordered.

Receiving and storing each product is a critical responsibility. Costly mistakes can come about from a staff member who was not properly trained in the correct procedures. A slight inaccuracy in an invoice or improper storing of a perishable item could cost the bakery hundreds of dollars.

Watch for a common area of internal theft. A collusion could develop between the delivery person and the employee receiving the products. Items checked as being received and accounted for may not have been delivered at all. Instead, driver simply keeps the items.

All products delivered to the bakery must be:

1. Checked against the actual order sheet.

2. The exact specification ordered (weight, size, quantity).
3. Checked against the invoice.
4. Accompanied by an invoice containing: current price, totals, date, company name and receiver's signature.
5. Weighed immediately to verify delivery amount..
6. Dated, rotated and put in the proper storage area immediately. Locked in their storage areas securely.
7. Noted for discrepancies on bills of lading, invoices and other related paperwork.

PLEASE NOTE: Vendors must be contacted immediately regarding overages and shortages. COD amounts should be revised to reflect the correct price when an error occurs. The delivery person must sign over the correction.

Keep an invoice box (a small mail box) in the kitchen to store all invoices and packing slips received during the day. Mount the box on the wall, away from work areas. Prior to leaving for the day, the receiver must bring the invoices to the manager's office and placed them in a designated spot. Extreme care must be taken to ensure that all invoices are handled correctly. A missing invoice will throw off the bookkeeping and financial records and statements.

Rotation Procedures

1. New items go to the back and on the bottom.
2. Older items move to the front and to the left.
3. In any part of the bakery: the first item used should always be the oldest.
4. Date and label all items.
   a. An office self-inking date stamp is handy.
   b. If more than one person is responsible for this task, they should have a distinct identifying mark or initial labels so you can identify who handled it.
   c. Dissolve-a-Way labels are handy for this task. Visit **http://www.daymarksafety.com/default.aspx** or call DayMark Food Safety Systems at 800-847-0101.

Issuing

Issuing ("checking out") raw ingredients is part of inventory control. All raw materials, from which portionable entrees are prepared or bulk ingredients for finished baked goods, must be issued on a daily basis. Whenever one of these bulk items is removed from a freezer or storage area, it must be signed out. Create a Sign-out Sheet that includes all items that you will issue to employees. The Sign-out Sheet should be on a clipboard near the storage unit.

When a part of a case or box is removed, the weight or quantity of the portion removed must be recorded. Once the item is signed out, the weight must be placed in the "Amount Used or Defrosted" column on the Daily Preparation Form (discussed next). This will show that the items signed out were actually used in the bakery. From this information, the kitchen director can compute a daily yield on each item prepared. This yield will show that the portions were weighed out accurately and the bulk product that was used to prepare baked goods and menu items. At any one of these steps, pilferage or waste can occur. The signing-out procedure will eliminate pilferage and help you create waste reduction procedures.

Products such as restroom or cleaning supplies may be issued in a similar manner. If these or other items were being stolen, the cost of each would show up in the cost projections at the end of the month.

Kitchen Controls

Combining a system of kitchen controls with your established procedures and policies will enable you to have an airtight food cost control system. The key to controlling food cost is *reconciliation*. Every step or action taken is checked and reconciled with another person. Management's responsibility is, once these systems are set up, to monitor them with daily involvement.

When all the steps and procedures are adhered to, you will know exactly where every dollar and ounce of food went. There are no loopholes. You and your management staff must be involved in the training and supervision of all employees. Daily involvement and communication is needed in order to succeed.

Employees must follow all procedures precisely. If they do not, they must be informed of their specific deviations from these procedures and assisted in correcting them. Any control initiated is only as good as the manager who follows up and enforces it. The time needed to oversee your controls shouldn't take more than an hour a day. This time is an investment in reducing waste, improving productivity and mentoring employees.

Simple manual systems can be created using predesigned forms that are worked with daily. Many of the basic purchasing and receiving functions are found in virtually all off-the-shelf accounting programs. See Chapter 13 for more information on computers and software for your bakery.

Daily Preparation

A Daily Preparation Form should be used by the journey bakers or prep cooks. At the end of each day, the head baker will list every item that must be prepared the following day. When the preparation staff arrives, they will count the number of items on hand from the previous night. As raw ingredients are issued to the prep cook, they will be recorded on the Daily Preparation Form.

As the day progresses, items will be prepared, dated, wrapped, rotated and placed in the racks for sale or delivery later that day. The piece count or number of portions prepared for each item is recorded in the "Amount Prepared" column. A sample form is below.

Amount Used (ingredients)	Amount Prepared	Beginning Amount	Day's Starting Total

The Amount Prepared plus the Beginning Amount equals the day's starting total. The starting total must be equal to or greater than the minimum amount needed. When all items are completed, the preparation sheet is placed in the manager's office.

Daily Yields

Daily yields represent the actual usage of a product from its raw purchased form to the prepared menu item. The yield percentage is a measure of how efficiently this was accomplished, or how effectively a preparation cook eliminated waste. The higher the yield percentage, the more usable material was obtained from that product.

All meat, seafood and poultry products must have a yield percentage computed for each entree every day. Yields are extremely important when determining menu prices. They are also very useful tools in controlling food cost. Daily yields should be computed by the kitchen director. An example of a daily **Yield Sheet** can be found at the end of this chapter.

Yield sheets should be kept for several months: they may become useful in analyzing other problem areas. All the information to compute each yield can be obtained from the Daily Preparation Form.

To Compute Yield Percentages

A. From the "Amount Issued" column, compute the total amount of ounces/pounds etc. used. Verify the amount in this column against the Sign-out Sheet. This figure is the starting weight in ounces/pounds.

B. The "Amount Prepared" column contains the number of units or portions yielded. Enter this figure on the Yield Sheet.

C. To compute the yield percentage, divide the Total Portion Weight (in ounces/pounds) by the Total Starting Weight (in ounces/pounds).

Yields should be consistent regardless of who prepares the item. If there is a substantial variance in the yield percentages (4–10 percent), consider these questions:

1. Are the preparation cooks carefully portioning all products? Over the months have they gotten lax in these methods?

2. Are you purchasing the same brands of the product? Different brands may have different yields.

3. Are all the items signed out on the Sign-out Sheet actually being used in preparing the menu items? Is it possible some of the product is being stolen after it is issued and before it is prepared? Do certain employees preparing the food items have consistently lower yields than others?

4. Is the staff properly trained in cutting, trimming and butchering the raw products? Do they know all the points of eliminating waste?

Periodically compare the average yield percentage to the percentage used in projecting the menu costs. If the average yield has dropped, you may need to review the menu prices.

Identifying Inventory Theft

Check the invoices every day for the items delivered that are in your inventory. Ensure that all items signed off as being delivered are actually in the storage areas. Should there be a discrepancy, check with the employee that signed the invoice. The number of items you start with (20) plus the number you received in deliveries (5), minus the amount signed out by the preparation cooks (1), must equal the number on hand (24). If there is a discrepancy, you may have a thief.

Chapter 20

Internal Bookkeeping:
Accounting For Sales and Costs

I nternal bookkeeping is where all financial transactions may be tallied, analyzed and reconciled. Your involvement in compiling, reporting and analyzing the financial data is critical to your ability to react quickly to problems, invest wisely and grow your business.

You should allot a couple of hours of each day to accounting. For many business people, hiring a skilled part-time bookkeeper is a wise investment. The bookkeeper's primary responsibility is to ensure that all sales, invoices and bills are accurately recorded and balanced. It is recommended that the bookkeeper not be used in any other capacity in the bakery, as he will be auditing the money and work of the other employees. The bookkeeper must understand and appreciate the confidential nature and importance of the work he is doing.

Your CPA should be used from time to time to audit the records, prepare financial and tax statements and lend management advisory services.

To locate a certified bookkeeper in your area, visit the American Institute of Professional Bookkeepers at **http://aipb.org**. If you cannot find an in-house bookkeeper, a neighborhood bookkeeping service can assist you.

Accounting Software

QuickBooks® by Intuit is a popular accounting package for the small business. Quick-Books (**www.quickbooks.com**) is rich in features including built-in remote-access capabilities and Web interfaces. Payroll and retail inventory modules are also available along with the coordinating tax program, TurboTax for individuals and sole proprietorships. Through Intuit, you can locate a local authorized QuickBooks consultant and accounting specialist to assist you with using this program.

Another popular accounting package is Sage 50 (available at **www.sage.com**). Sage 50, like QuickBooks, offers an assortment of features and more than one version.

There are also QuickBooks and Peachtree certified accountants and bookkeepers that can assist you with setting up your accounting system and work directly with your files. An online consulting firm also offers information on QuickBooks for Restaurants at **www.rrgconsulting.com**.

If you are just setting up your accounting program and decide to use QuickBooks®, an add-on product called "The Tasty Profits Guide to QuickBooks® Software for Bakeries" is available. This helpful guide to QuickBooks® enables you to save thousands of dollars doing your own accounting with its proven, easy-to-use system. Simply install the floppy disc that is included with the "Tasty Profits Guide" directly into your computer. You'll download the preconfigured bakery accounts, and you are ready to go. You will have instant access to all your financial data; calculate accurate food and bar costs with ease; reconcile bank and credit-card statements; track and pay tips that are charged to credit cards; and calculate sales tax automatically. The program costs about $70 and is available at **www.atlantic-pub.com** and 800-541-1336: Item TP-01.

TwinPeaks Software offers bakery management software (including accounting) and a free guide to selecting accounting software at **www.twinpeaks.net/twinpeaks/free-stuff2.html**.

Your accountant can also give you their recommendations or you can check the free service by CPA Online at **www.findaccountingsoftware.com**. In addition, several of the bakery management software packages above offer accounting (accounts receivable, accounts payable and inventory) components. Additional information on computers and software can be found in Chapter 13.

Managing the Bakery's Cash Flow

Daily involvement and analysis of your financial records are necessary if the bakery is to take full advantage of the credit terms and discounts offered by suppliers. Substantial savings can be acquired simply by managing the bakery's cash flow and utilizing its purchasing power.

After you've been operating for a few months, most purveyors will extend 30-day net terms if you request them. This is an advantageous situation; through proper management, the bakery's inventory may be turned over as many as five or six times in a 30-day period. In effect, the purveyors will be financing your operations. Few businesses can turn their inventories over this quickly, so they are forced to pay interest or finance charges. Quick turnover is one of the blessings of the bakery business. Careful planning and synchronization between the purchasing and bookkeeping departments are needed to obtain maximum utilization of the cash flow. The savings are well worth the additional effort.

CODs and Your Cash Flow

It is not uncommon for vendors to require payment upon delivery (this is especially true for new businesses). The best tracking procedure is to have a check ready. However, small deliveries can be paid "from the till" by an authorized employee. You'll need to take extra precautions with this procedure. A memo form should be filled out with the particulars and placed in the till so that the assigned cashier can properly balance at the end of their shift. Employees must be extra careful when accepting COD deliveries because your "leverage" in problem solving and the sense of urgency is diminished with a vendor who has already been paid.

Cash Handling

Proper cash handling procedures are critical. Below are some helpful cash handling, sales reporting and auditing procedural tips.

1. If possible, individual cash drawers should be available for each cashier. This will help you determine if someone isn't handling the duty accurately. Computerized systems can have specific coded keys to keep tallies separate.

2. Cash drawers should be prepared and assigned each morning. The responsible cashier should recount the drawer's contents and sign a receipt for the funds.

3. At the end of each shift, the responsible cashier should count and balance their drawer against the cash register's report.

4. All overages or shortages should be written up and signed off by a supervisor. Drawers are "checked" back into the safe with the responsible supervisor.

5. Checks and credit card receipts should be tallied separately and stored in the safe until the bookkeeper is ready to deposit them.

6. Automated credit card processing can be a real time saver and you'll receive the funds faster.

7. For "made to order" items, use sequentially numbered guest checks and/or custom product forms. Maintain control over these to monitor potential employee fraud. Voided guest checks should be kept and turned in at the end of the day.

8. For phone or Web orders, make certain the credit card's 4-digit ID code is listed.

9. For in-person purchases, write the customer's phone number on the form. If an error is made, you'll be able to contact the customer

10. If a tip was charged, a "Cash Paid Out" from the cashier/server should be given to the employee. The "Cash Paid Out" is not a purchase, because when the charge goes through, the bakery will be reimbursed.

    a. Set up a cash reserve or special account and reimburse the cash drawer for this "Cash Paid Out." When the check from the credit-card company comes in, put the "Paid Out" amount back into the reserve or special account.

Sales Categories

To compute individual category percentages, divide the category sales by the total daily sales. "Actual Month-to-Date Sales" is a tally of the daily sales. "Cash, Over/

Short" refers to any mistakes made at the register. Complimentary, house and manager figures must also be recorded as no charge sales.

Every item and sale is accounted for and reconciled against every other transaction in the bakery. Keep all of these forms for at least five years in a fireproof storage file. All forms used during the month may be kept in loose-leaf binders in the bookkeeper's office. If you are using a computerized system, make daily backups and keep copies of monthly backups off site.

Payroll

Getting your employees paid and withholding deposits made accurately and on time can be handled by an in-house bookkeeper with a computerized payroll program or accounting module (such as QuickBooks® or Peachtree®). However, you may find it more convenient to use a your accountant, a local payroll service, bank provided services or a Web-based service.

Although you may decide to use an outside consultant, your bookkeeper will still be involved in the computation of the daily labor costs. After each pay period, the bookkeeper will need to compute each employee's time card and call the information to the payroll service company or key the info into the accounting software. There are time clocks now available that can link employee scheduling, time clock administration and accounting all into one foolproof system.

Your employees all work for your bakery; however, you may find it advantageous to calculate their hours and associate them with specific bakery activities and/or profit centers. For example: 6 hours baking bread and 2 hours handling front counter relief.

The manager and owner salaries should be listed separately at the bottom of the Payroll Form. These costs are separated, as they are budgeted differently.

The month-to-date payroll percentage is computed by dividing month-to-date sales by the month-to-date actual payroll costs. The budget figures are the budgeted total labor costs divided by the number of days in the month. The month-to-date payroll column is the prorated budgeted amount.

Tipped Employees

One of the biggest challenges facing food service operators and managers in regard to payroll is getting employees to report and pay taxes on their tips, as required by the IRS. This can be especially difficult in environments where tipping is less common than in full-service restaurants. Complying with the intricacies of the tip reporting and allocation rules can be difficult and confusing and the applicable laws are constantly changing.

You must use extreme caution in this area as the IRS can hold you liable for taxes relating to tips. It is recommended that you review the issue with your accountant and/or attorney. Up to date information on food service tips is available from the IRS (**www.irs.gov**), your state restaurant and bakery associations and the National Restaurant Association (**www.restaurant.org**).

Employer –Tip Records

It is in your company's best interest to insist that all employees accurately report their income from tips. The IRS will hold you responsible. Establishments that do not comply are subject to IRS audit and possible tax liabilities, penalties and interest payments. As a precaution, if you have any employees who customarily receive tips from customers, patrons or other third parties, we recommend you keep the following additional information about tipped employees:

- Indicate on the pay records – by a symbol, letter or other notation placed next to his or her name - each tipped employee.

- Weekly or monthly amount of tips reported by each employee.

- The amount by which the wages of each tipped employee have been increased by tips.

- The hours worked each workday in any occupation in which the employee does not receive tips, and the total daily or weekly earnings for those times.

- The hours worked each workday in any occupation in which the employee receives tips, and the total daily or weekly straight-time earnings for those times.

Large Food or Beverage Establishments Need to File Form 8027 with the IRS. You may meet the definition of a "large food or beverage establishment" if you employ more than 10 employees. If you do, the law requires that you file Form 8027, *Employer's Annual Information Return of Tip Income and Allocated Tips*, with the IRS.

If you meet the definition, the law requires that you report certain tip information to the IRS on an annual basis. You should use Form 8027 to report information such as total charged tips, charged receipts, total reported tips by employees and gross receipts from food-and-beverage operations. Also, employers must allocate tips to certain directly tipped employees and include the allocation on their employees' W-2 forms when the total of reported tips is less than 8 percent.

The Tip Rate Determination and Education Program

The Tip Rate Determination and Education Program was developed by the Internal Revenue Service in 1993 to address the concern of widespread underreporting of tip income in the food-and-beverage industry. The goal was to involve employers in monitoring their employees' tip-reporting practices.

There are two different IRS programs available: the Tip Rate Determination Agreement (TRDA) and the Tip Reporting Alternative Commitment (TRAC). Participation in one of these programs is voluntary, and the bakery may only enter into one of the agreements at a time. *Please note that 1998 tax legislation specifies that IRS agents can't threaten to audit you in order to convince you to sign a TRAC agreement.*

The big benefit for you as an employer is that you will not be subject to unplanned tax liabilities. Those who sign a TRAC or a TRDA agreement receive a commitment from the IRS that the agency will not examine the owner's books to search for under-withheld or underpaid payroll taxes on tip income. There are benefits to employees, also, including increases in their Social Security, unemployment compensation, retirement plan and worker's compensation benefits.

Under TRDA, the IRS works with you to arrive at a tip rate for your employees. Then at least 75 percent of your tipped workers must agree in writing to report tips at the agreed-upon rate. If they fail to do so, you are required to turn them in to the IRS. If you do not comply, the agreement is terminated and your business becomes subject to IRS auditing.

The TRAC is less strict but requires more work on your part. There is no established tip rate, but you are required to work with employees to make sure they understand their tip-reporting obligations. You must set up a process to receive employees' cash tip reports, and they must be informed of the tips you are recording from credit-card receipts.

Tip Credits for Employers

As an employer, you may also be eligible for credit for taxes paid on certain employee tips (IRS Form 8846). The credit is generally equal to the employer's portion of Social Security and Medicare taxes paid on tips received by employees. You will not get credit for your part of Social Security and Medicare taxes on those tips that used to meet the federal minimum wage rate applicable to the employee under the Fair Labor Standards Act. This is also subject to state laws. You must also increase the amount of your taxable income by the amount of the tip credit.

More Information on Tip Reporting

The following IRS forms and publications relating to tip income reporting can be downloaded directly from the government Web site: **http://apps.irs.gov/app/ picklist/list/formsPublications.html**.

- Pub 505 – Tax Withholding and Estimated Tax
- Pub 531 – Reporting Tip Income
- Form 941 – Employer's Quarterly Federal Tax Return
- Form 4137 – Social Security and Medicare Tax on Unreported Tip Income
- Form 8027— – Annual Information on Tip Income and Allocated Tips
- IRS Guide P1872for – Tips on Tips, a Guide to Tip Income Reporting for Employees in the Food and Beverage Industry (This is an excellent 8-page hand-out.)
- IRS Guide P1875 – A Guide to Tip Income Reporting for Employers in the Food and Beverage Industry

Chapter 21

Successful Budgeting
and Operational Management

T o be financially successful, you must set up a long-range plan detailing how much money you want the bakery to return and when. Budgeting is an accounting record and a tool used to evaluate how effectively the bakery, management and employees performed during the month. Reviewing this information will help you recognize cost problem areas and act quickly to correct them. We recommend you use a monthly budget because of the fluctuating operating performances common in food service establishments.

Monthly Budgeting

Once set up and operating, about four hours each month is all that will be required to compute the old budget and project a new one. Although the bakery may only be in the pre-opening stage, it is imperative that you start to develop an operating budget now. As soon as the budget is prepared, you will possess the control for guiding the business towards your financial goal.

There are many other benefits to preparing and adhering to a monthly operational budget.

- Supervisors and key employees will develop increased awareness and concern about the bakery and controlling its costs.

- A well-structured, defined budget and orderly financial records will aid you greatly in obtaining loans and will develop an important store of information should you decide to expand or sell in the future.

- Your financial decisions and forecasts will become increasingly consistent and accurate, as more information will be available to you. Projecting the Operational Budget

Projections often frustrate new business owners. However, even if your projections aren't initially accurate you'll be gathering and comparing valuable data – and you'll get better at it! The only way to start is to jump right in.

Total Sales

Projecting total sales is the most crucial and difficult aspect of budgeting. The fact that it is impossible to know how business will be from day to day makes budgeting total sales a perplexing task. Most costs are either variable or semi-variable, which means they will fluctuate directly in relation to the total monthly sales. Thus accurately projecting these costs depends largely upon using an accurate total sales figure. Projecting total sales, at first, will be difficult—and most likely inaccurate—but after several months of operation, your projections will be right on target. You will be surprised at how consistent sales and customer counts are and how easy it will be to consistently budget accurately.

The initial budgets may be unrealistic expectations. Sales will probably be low, as you will not have been able to build a substantial clientele or reputation. Operating costs will be higher than normal. It will take a couple of months to streamline and build an efficient bakery even with the best-laid plans. All of these costs are normal and should be anticipated. Profit margins will be small and possibly even nonexistent.

During your start up (or when introducing a new product or launching a new service) allow 4–12 weeks to ensure that your products are perfected and all the bugs are worked out of the systems. This is no time to cut back on costs. Your intention is to be in business for a long time. Allocate sufficient funds now to make sure the business gets off on the right foot and profits will be guaranteed for many years.

Schedule a full staff every day to make certain all details will be covered. Discontinue those items that are passable but not of the quality level desired. Slow, clumsy service and only average food will never build sales. Strive for A-1 quality products and service. Constantly reiterate to employees this primary concern and before long they will self-monitor the quality. Once you develop a clientele and a solid reputation for serving consistent, quality products, the budget and profits will fall into place.

Tracking Sales Growth

If you have purchased an existing bakery, determine the historical sales average by month. This will be your benchmark data. You'll also be able to see a cycle to the sales volume. Perhaps December is a booming month, but July will be dreadfully slow. This information will help you with staff scheduling and purchasing.

Restaurants typically use customer counts (each individual diner) in calculations and comparisons. If your accounting separates your baked goods retail from your prepared meals (take-out or dine-in), you will be able to use these food-service ratios. For information on these calculations, check out Virtual Restaurant at **www. virtualrestaurant.com**. Try working with actual sales totals and using customer counts to determine which formula (or perhaps both) provide you with the most accurate assessment.

When computing and analyzing your bakery activities, keep in mind that each period of time must have the same number and type of days. In other words, you can only accurately compare months that have the same number of Mondays, Tuesdays and so forth, since sales are different for each day; the results and analysis would be inaccurate. The most accurate way to analyze the percentage of growth or loss is to compare the previous month to the same month last year and then compare the percentage to the current month. Remember to only examine actual cover counts as indicators of growth; changes in sales may be the result of a price change.

Review and analyze the growth in sales volume during the past year, current year and current month. Based upon past sales figures, determine the percentage of growth or decline in growth anticipated in the coming month. Percentage of growth or decline can be computed by subtracting the most recent period of by the past period.

Calculating Raw Material Costs

Just like any manufacturer, you use raw materials that you transform into finished goods. Calculating your raw material costs is the first step to determining your sales price. These costs are variable as the more you make and sell – the greater the total costs.

Food Costs

Food costs for take-out and dine-in meals are calculated using the cost-of-sales percentage used in the restaurant industry. Food costs will also fluctuate directly with the sales variance. Just as with raw materials, the more food and beverage you sell; the higher your food costs. In budgeting material costs, the important figure to analyze is not the actual cost but the percentage of cost, or the *cost-of-sales percentage*, as it is more commonly known.

Compute the cost-of-sales percentage by dividing the actual cost of the category by the category's total sales. The result will be a percentage figure. This formula will present an accurate indication of the category's costs, as the cost of sales are proportionate to each other.

The cost-of-sales percentages for each category – food, beverage and other – are used to determine the selling price of each item. Food cost was projected at between 35 and 40 percent cost of sales. Enter the percentage figure used in the previous sections in the "Budgeting Percentage" column. This percentage figure can be used initially in order to project the first month's budget. After several months of operation, the actual figure can be substituted.

Multiply the individual material costs by each respective budgeted percentage; the results are the budgeted cost amounts. For example: if food sales were budgeted at $100,000, and the food cost percentage was estimated at 40 percent, the budgeted food cost would be $40,000.

Increases in product costs will raise the actual cost-of-sales amount; adjust the budgeted amount accordingly. However, be certain that if an increase is anticipated, the increase will affect the following month, which is what is being budgeted. Items purchased at a higher price and then stored in inventory will have no effect upon the following months actual cost of sales, as the product will not have been used. Add all three budgeted costs to compute the total budgeted material costs. Subtract the total gross costs from the total sales to compute the gross profit dollar amount. Divide gross profit by total sales to calculate the gross profit percentage.

LABOR

Manager Salary

Manager salaries should be a fixed monthly cost. Total all the manager salaries for one year; divide this figure by the number of days in the year (usually 365), and multiply this cost by the number of days in one month. Salary changes during the year will require adjustments. When owners take an active part in the management of the bakery, or when the company is incorporated, the owners should have their salary amount included in this category.

Employee Salary

The employee salary expense is a semi-variable cost which will fluctuate directly with total sales. Employee labor costs have a breakeven point, the point where the labor cost is covered by the profit from sales. As this point is reached and total sales increase, the labor cost percentage will decrease, increasing net profit. Thus, the cost of labor is determined by its efficiency and by the volume of sales it produces. Multiply the projected total sales by the average labor cost percentage to arrive at the anticipated labor cost dollar amount. Adjust this figure in relation to the amount of employee training anticipated for the month.

Overtime

Overtime should be nonexistent – or at least kept to an absolute minimum. No amount should be budgeted for overtime. Money spent on overtime usually indicates poor management and inefficiency. Bookkeepers should be on the lookout for employees approaching 40 hours of work near the end of the week. Carefully prepared schedules will eliminate 98 percent of all overtime work and pay. Employees who wish to switch their schedules around should only be allowed to do so after approval from the manager.

Controllable Operational Costs

Dishware and Utensils

You'll have a wide variety of serving utensils and perhaps even some dishware for dining in. If you are encouraging regulars to stop by frequently, you may find that dishware (as opposed to disposables) makes them feel more at home.

Capital Expenditures

Capital expenditures for equipment, vehicles, computers and software can be depreciated over a set lifespan or under current IRS regulations may be deducted entirely the year it is purchased. After your launch your business, these costs can be spread out through financing or leasing agreements.

Kitchen and Dining Room Supplies

These are consumable items that need to be replaced frequently. Proper inventory procedures will help you keep these costs down.

Office Supplies

Cost of office supplies should be a fixed dollar amount each month. Capital expenditures must be depreciated.

Uniforms

The uniform expense will depend upon the state in which the bakery is located and individual management policies. Some states allow the company to charge the employees for uniforms; others do not. Many bakeries that do charge employees for uniforms do so at cost, which, if done correctly, should cost the bakery nothing but administrative time.

Laundry and Linen

Laundry and linen purchases should be a consistent monthly expenditure, as laundry and linen is usually bought once or twice a year, in bulk. This expenditure column is for the purchase price only; cleaning is computed in a separate column, under "Services."

Services

Laundry Cleaning

Cleaning of laundry is variable expense directly related to total sales. Multiply last month's percentage of cost by budgeted sales. Adjust the figure for price increases.

Security

Security should be a consistent, fixed monthly expenditure. Service-call charges should be coded to "Equipment Repairs" under "General Operating Costs."

Freight

Incoming freight charges are usually included in the purchase price. However, you may book charges from an independent carrier for delivery costs.

Legal

Legal service is variable expense that can fluctuate greatly. Estimates for most legal work can be obtained, but it's best to budget a little each month to cover periodically large legal fees.

Accounting

A semi-fixed expense depending upon the amount and the type of accounting services used. Once set up and operating, the accounting expense should be a consistent monthly charge except for an annual tax-preparation and year-end audit fee.

Maintenance

Facility maintenance should be a fixed monthly expenditure if using a maintenance service company with contract service. This may also include parking lot cleaning, window cleaning and other periodic cleaning services.

Payroll

This is a semi-fixed expense fluctuating directly with the number of employees on the payroll. Bakeries not utilizing a computerized payroll service will not have a payroll preparation expense. The wages paid to the bookkeeper are included in the employee labor expenditure.

Utilities

Telephone

Telephone service should be a relatively consistent monthly expense. All long-distance phone calls should be recorded in a notebook (Your local office-supply store has a specially designed book for this purpose.) The itemized phone bill should be compared against the recorded phone calls to justify each one.

Water

Water should be a semi-variable expense.

Gas

Gas may be a variable or semi-variable expense depending upon the type of equipment it operates. Gas used in heating will be a variable expense, because more will be used during the winter months than in the summer.

Electricity

Electricity may be a variable or semi-variable expense depending upon the type of equipment it operates. Electricity bills are normally higher during the summer months, as this is when the air-conditioning units are used.

Heat

Heat includes the cost of any heating material used but not listed above, such as coal, wood, oil, etc.

Fixed Operating Costs

Rent

This should be monthly amount of rent or, if the building is leased, the monthly lease. Certain business-rental and lease agreements also include payment of a percentage of the total sales or per-tax profit amount. Should this be the situation, use the budgeted total sales figure and project the anticipated amount due. Enter this amount and the total rent amount in the "Budgeted" column.

Insurance

Total all insurance premium amounts (fire, theft, liability, worker's compensation, etc.) and divide by 12. This figure will equal the average monthly insurance expense.

Property Taxes

If applicable, divide the annual property tax amount by 12. This figure will equal the average monthly property tax amount.

General Operating Costs

Labor Taxes

This is the tax amount the employer is required to contribute to the state and federal government. A separate tax account should be set up with your bank to keep all the tax money separate. Labor taxes include: Social Security, federal unemployment tax and state unemployment tax.

Other Taxes

This includes all miscellaneous taxes, such as local taxes, sales tax paid on purchases, etc. This column is for any tax the bakery pays for goods and services. It is not for sales tax or other taxes the bakery collects, as they are not expenditures. Federal income tax is not a deductible expenditure and should not be listed here either.

Repairs: Equipment

This includes the cost of scheduled and emergency repairs and maintenance to all equipment. Always budget a base amount for normal service. Adjust this figure if major repairs or overhauls are anticipated.

Repairs: Building

This includes the cost of minor scheduled and emergency repairs and maintenance to the building. Always budget a base amount for normal repairs and maintenance. Large remodeling or rebuilding projects should be budgeted as a separate expenditure and depreciated.

Entertainment

Entertainment expenses are deductible only if the amounts spent are directly related to the active conduct of the business.

Advertising

Advertising includes all the costs of advertising the bakery, including television, radio, mailing circulars, newspapers, etc.

Promotional Expense

This is the expense of promotional items: key chains, calendars, pens, free dinners, T-shirts, sponsorship of sporting events, etc.

Equipment Rental

This cost is the expense of either short- or long-term renting of pieces of equipment or machinery.

Postage

This is postage paid for business purposes.

Contributions

These are all contributions paid to recognized charitable organizations.

Trade Dues, Business Associations

This includes dues paid to professional organizations such as the National Bakery Association and business organizations such as the Better Business Bureau. Trade magazine subscriptions should also be entered in this category. This expense should be divided by 12 to apportion the cost from the current month.

Licenses

This is the expense of all business and government licenses: operating licenses, a health permit, liquor licenses, etc. This expense should also be divided by 12 to apportion the cost from the current month.

Credit Card Expense

Credit card expense can be computed by multiplying the service-charge cost-of-sales percentage by the total projected credit-card sales volume.

Travel

Travel includes the expense of ordinary and necessary travel for business purposes for yourself and your employees. Expenses incurred with delivery service would be booked under vehicle expense.

Bad Debt

This expense should be nonexistent if the proper procedures for handling credit cards and checks are enforced. Normally, the full amount of a bad debt is a tax-deductible expense. However, you must prove the debt is worthless and uncollectible. In some states, the employee who handled the transaction may be legally held liable for the unpaid amount.

Total Net Profit

Subtract "Total Budgeted Expenditures" from "Total Sales." The result is the total net profit (or loss). Divide the total net projected profit by projected "Total Sales" to compute the projected "Pre-Tax Net Profit Percentage." Total projected sales minus total material costs will equal the gross profit amount.

Depreciation

Depreciation may be defined as the expense derived from the expiration of a capital assets quantity of usefulness over the life of the property. Capital assets are those assets that have utility or usefulness of more than one year. Since a capital asset will provide utility over several years, the deductible cost of the asset must be spread out over its useful life—over a specified recovery period. The IRS publishes guidelines for the number of years to be used for computing an asset's useful life. Each year a portion of the asset's cost may be deducted as an expense. Your accountant can advise you on the rules and benefits of immediate deductibility of equipment and vehicles as compared to depreciating these.

Some examples of depreciable items commonly found in a bakery include: office equipment, kitchen and dining-room equipment, the building (if owned), machinery, display cases and any intangible property which has a useful life of more than one year. Thus, items such as light bulbs, china, stationery and merchandise inventories may not be depreciated. The cost of franchise rights is usually a depreciable expense.

Successful Managing of Operational Costs and Supplies

When establishing your operational budget, you divided supplies into various categories.. Creating an *Operational Ordering Form* and an *Operational Inventory Form* will help you keep track of daily activity. Whenever a new product is ordered, enter the new item on both forms. When it comes time to do the weekly order and monthly inventory you will not miss or forget anything, as all the items will be listed. Certain items may fit into two categories because they are used in several areas of the bakery. Place the item in the category where it is used the most. This will not affect the cost projection as long as the item is listed in only one category.

Ordering Operational Supplies

Ordering operational supplies must be carefully thought out. Too large of an inventory (back stock) will tie up operating capital unnecessarily. However, you must have enough inventory to cover the unexpected. Having an employee run out for toilet paper isn't cost effective. Your inventory levels determined by the speed orders can arrive, the quantity discounts and your typical usage from delivery to delivery.

Portion control items such as scales, scoops and ladles are often difficult to obtain; when you find a supplier that has the size and type you desire, order several and keep them in reserve. A common, but poor, excuse for not portion-controlling products is that the employee does not have access to the right type of utensil. Your responsibility is to provide the employees with proper tools to do their jobs.

Beginning Inventory

Computing this figure is similar to computing the beginning inventory amount for food and beverage. However, there is one difference. The beginning inventory amount for each operational category is the dollar amount that is *in storage* when the bakery is totally set up. The reason for this is that operational supplies are projected for each month. When you first open, the cost of setting up the bakery is considered a one-time start-up cost. Operational costs are thus a measure of how well you controlled these costs *following* start-up. Separating this start-up cost may have some additional tax advantages. Your accountant will be able to advise you on this possibility.

Chapter 22

Basic Cost Control For Food Service Operations

Cost Controls Are Crucial

Profitability in a labor-intensive bakery is directly related to your ability to control your costs for food, overhead and people. Cost control is about collecting, organizing, interpreting and comparing numbers that affect your bottom line and tell you the *real story* of what's going on in your bakery. This is not a job that can be easily delegated, because these numbers are valuable decision-making tools.  Working directly with these figures gives you the ability to react quickly to improve your control methods, purchasing procedures and employee training.

Your first step is to create systems and procedures to gather data. This data will be the foundation for analyzing your labor productivity, portion control, purchase prices, marketing promotions, new products and competitive strategy.

During your first few months of operation, your data may not be precise; however, you are establishing benchmarks to gauge future activities. If you have purchased an existing bakery, you should have records from the prior owner that will be your benchmark. Franchise operations will be able to give you data based on historical information from other operations in similar communities.

Profits May Hide Problems

Business owners reaping high profits should not be lax on monitoring costs. High profits can hide inefficiencies that will surely expose themselves during times of low sales. Many people become cost-control converts only *after* suffering losses. This is shortsighted. The idea of prevention versus correction is fundamental. Prevention occurs through advanced planning. Your primary job is not to put out fires, it's to prevent them—and to maximize profits in the process.

Controlling Costs Works

Controlling costs works – all the time – because it focuses on getting *the most value from the least cost* in every aspect of your operation. By keeping costs under control, you can charge less than the competition or make more money from charging the same price.

There are operating advantages and opportunities that are not afforded you if you don't know what you're spending. And if you don't know what you are spending, you cannot control it. Furthermore, most of the waste that occurs in bakeries can't be detected by the naked eye. It takes records and reports—whose meanings you've mastered interpreting—to tell you the size of the inefficiencies that are taking place.

The larger the distance between an owner or manager and the actual daily bakery operation, the greater the need for effective cost control monitoring. This is how chain bakeries and bakery franchisors keep their eyes on thousands of units around the world.

Cost control is not accounting or bookkeeping: these are the information-gathering tools of cost control. Cost control can be defined by explaining its purposes:

- To provide management with information needed for making day-to-day operational decisions.
- To monitor department and individual efficiency.
- To inform management of expenses being incurred and incomes received and whether they fall within standards and budgets.
- To prevent fraud and theft.
- To provide the ground for the business's goals (not for discovering where it has been).
- To emphasize prevention, not correction.
- To maximize profits, not minimize losses.

Cost control is not a sign of distrust towards their staff. The main purpose of cost control is to provide information about daily operations. Prevention of theft is a secondary function. Cost controls are about knowing how you got to where you are and where you are going.

Interpreting the Numbers

Understanding those numbers means interpreting them. To do this effectively, you need to understand the difference between control and reduction. Control is achieved through the assembly and interpretation of data and ratios on your revenue and expenses. Reduction is the actual action taken to bring costs within your predetermined standards. Effective cost control starts at the top of an organization. Management must establish, support and enforce its standards and procedures.

There are 10 primary areas that are central to any food production/service operation and are therefore crucial elements of cost control records:

1. **Purchasing.** Your inventory system is the critical component of purchasing. Before placing an order with a supplier, you need to know what you have on hand and how much will be used in a given time.
   - Allow for a cushion of inventory so you won't run out between deliveries.
   - Compare cost per unit of non-perishables vs. usage to determine whether volume buying makes fiscal sense.
   - Once purchasing has been standardized, the manager simply orders from your suppliers.
   - Records show supplier, prices, unit of purchase, product specifications, etc. This information needs to be kept on paper.

2. **Receiving.** This is how you verify that everything you ordered has arrived.
   - Check for correct brands, grades, varieties, quantities, correct prices, etc. Products purchased by weight or count need to be checked.
   - Incorrect receivables need to be noted and either returned or credited to your account.

3. **Storage.** Proper storage, with regard to temperature, ventilation and freedom from contamination, ensures food products remain in optimum condition until being used.
   - Expensive items need to be guarded from theft.

- Well-organized storage will reduce the time it takes to inventory items and minimize "assuming" you have plenty on hand.

4. **Issuing.** Procedures for removing inventory from storage are part of the cost control process.
   - Head bakers and counter managers have authority to take or "issue" stock from storage to the appropriate place. This is a much more important aspect of cost control than it seems, because without this data you can't determine accurate sales figures.
   - To calculate your cost of goods sold, you need to know your beginning inventory, how much was sold and your ending inventory.

5. **Rough Preparation.** How your staff minimizes waste during the preliminary processing of inventory is critical.

6. **Preparation for Sale/Service.** Packaging of finished baked goods or components of menu items.
   - Does the packaging create waste or "defects"?
   - Can procedures be improved to minimize waste?

7. **Portioning/Transfer.** Food can be lost through over-portioning. For example, cream cheese on toasted bagels. Final preparation should be monitored regularly to ensure quality and quantity standards are being adhered to. This is such a crucial element to cost control.

8. **Order Taking.** Every item sold or issued from the kitchen needs to be recorded by computer, cash register or manually.
   - Basically, it needs to be impossible for anyone to get food or drinks without having them entered into the system. Reinforce that this isn't a distrust procedure but a profit analysis procedure.
     1. Allow for discretionary "give aways" for customer service purposes.
        - Provide employee meals, snacks and beverages as a benefit – but keep track of these. (For info on deducting food-service employee meals, read IRS Publication 15-B).
   - No verbal orders should be accepted by or from anybody—including management and owners.

9. **Cash Receipts.** Monitoring sales is crucial to cost controls. Under/over-charging and lost paperwork for custom and will call orders must be

investigated daily. Sales information must be compiled to build a historical financial record. This record helps you forecast the future.

10. **Bank Deposits/Accounts Payable.** Proper auditing of bank deposits and charge slips must be conducted.

Cost control is an ongoing process that involves every employee. Setting easy-to-use systems and procedures in place and explaining the "why" behind the "because" will make your data collection more accurate. A continuous appraisal of this process is integral to the functioning of your bakery. There are five key elements to an effective cost control strategy:

1. Planning in advance.
2. Procedures and devices that aid the control process.
3. Implementation of your monitoring and improvement programs.
4. Employee compliance.
5. Management's ongoing enforcement and reassessment.

Furthermore, your program should be assessed with the following questions:

*   Do your cost controls provide relevant information?
*   Is the information timely?
*   Is it easily assembled, organized and interpreted?
*   Are the benefits and savings greater than the cost of the controls?

Penny Wise, Pound Foolish

This last point is especially important. When the expense of the controls exceeds the savings, that's waste, not control. Spending $30,000 on a computer system that will save you $5,000 in undercharges is ineffective.

Setting Standards

Standards are key to any cost control program. Predetermined points of comparison must be set, against which you will measure your actual results. The difference between planned resources and resources actually used is the variance. Management can then monitor for negative or positive variances between standards and actual performance and will know where specifically to make corrections. These five steps illustrate the uses of standards:

*   Performance standards should be established for all individuals
    and departments.

- Individuals must see it as the responsibility of each to prevent waste and inefficiency.
- Adherence—or lack of adherence—to standards must be monitored.
- Actual performance must be compared against established standards.
- When deviations from standards are discovered, appropriate action must be taken.

Your job is to make sure standards are adhered to. Is your staff using measuring scoops and ladles and sized bowls, glasses and cups, weighing portions individually, portioning by count, and pre-portioning? These are all useful tools to make sure standards are met and your cost control program implemented effectively.

Cost Ratios

Owners and managers need to be on the same page in terms of the meaning and calculation of the many ratios used to analyze food, beverage, and labor costs. It's important to understand how your ratios are being calculated, so you can get a true indication of the cost or profit activity in your bakery. Numerous cost control software programs are available with built-in formulas for calculating ratios and percentages.

The Uniform System of Accounts for Restaurants (USAR), published by the National Restaurant Association (www.restaurant.org), is an essential guide for food service accounting. It establishes a common industry language that allows you to compare ratios and percentages across industry lines. The goal of this comparison is to create financial statements that are management tools, not just IRS reports. Cost control is not just the calculation of these numbers. It's the interpretation of them and the appropriate (re)actions taken to bring your numbers within set standards.

Bakeries are actually "manufacturers" and would employ production cost ratios and analysis. If you also a café or deli and offer "prepared" food for dine-in or take-out, you'll also use "restaurant" cost ratios and analysis. Your style of service may be closer to a "fast food" environment so your review of data should take this into consideration.

Food Cost Percentage

This basic ratio is often misinterpreted because it can be calculated so many ways. Basically, it is *food cost* divided by *food sales*. However, whether your food cost is determined by food *sold* or *consumed* is a crucial difference. In addition for your food cost percentage to be accurate, a month-end inventory must be taken. Without this figure, your food cost statement is basically useless. This is because your inventory will vary month to month – even in the most stable environment (which yours probably won't be initially) – because months end on different days of the week.

Distinguishing between food *sold* and *consumed* is important because all food consumed is not sold. Food consumed includes all food used, sold, wasted, stolen or given away to customers and employees. Food *sold* is food bought at full price.

Maximum allowable food cost (MFC) is the most food can cost you to prepare and service while still making a profit. If at the end of the month, your food cost percentage is over your maximum allowable, you won't meet your profit expectations. This is how you calculate it:

1. Total your monthly costs (Write your dollar amounts of labor costs and overhead expenses and exclude food costs. Refer to past accounting periods and yearly averages to get realistic cost estimates.

2. Add your monthly profit goal as either a dollar amount or a percentage of sales.

3. Convert dollar values of expenses to percentages by dividing by food sales for the periods used for expenses. *Generally, don't use your highest or lowest sales figures for calculating your operating expenses.* Subtract the total of the percentages from 100%. The remainder is your maximum allowable food cost percentage (MFC).

    100% - (Total monthly costs – food costs + monthly profit goal) = MFC %.

Actual food cost percentage (AFC) is the percentage you're actually operating at. It's calculated by dividing food cost by food sales. If you are deducting employee meals from your income statement, then you are calculating cost of food *sold*. If there is no deduction of employee meals—which is true for most operations—then the food cost you're reading is food *consumed*. This is always a higher cost than food sold, and if inventory is not being taken, the food cost on your income statement is just an estimate based on purchases and isn't accurate.

Potential food cost percentage (PFC) is also called your theoretical food cost. This is the lowest your food cost can be because it assumes that all food consumed is sold, and that there is no waste whatsoever. It is found by multiplying the number sold of each menu item by the ideal recipe cost.

Standard food cost (SFC) is how you adjust for the unrealistically low PFC. This percentage includes unavoidable waste, employee meals, etc. This food cost percentage is compared to the AFC, and is the standard management must meet.

Prime food cost includes with the food cost the cost of direct labor. This is labor incurred because the item is made from scratch—baking pies and bread, trimming steaks, etc. When the food cost is determined for these items, the cost of the labor needed to prepare them is added. So prime cost is food cost plus necessary direct labor. This costing method is applied to every menu item needing extensive direct labor before it is served to the customer. Indirect labor cannot be attributed to any particular menu item, and is therefore overhead. Prime cost is the total cost of food and beverage sold, payroll, and employee benefits costs.

Beverage Sales

Coffee, tea, milk or juices are typically considered food sales in food service establishments. However, in a retail bakery environment, you may want to consider these as a separate profit center and monitor beverage sales separately. This can be beneficial, for example, if you are interested in investing in expensive cappuccino equipment. Also remember that wherever you include soft drinks, your food costs will decline because the ratio of cost to selling price is so low.

Inventory

Inventory turnover is calculated by dividing cost of food consumed by your average inventory (This is simply your beginning inventory plus your ending inventory, divided by 2).

Check Average

Check average is not just total food and beverage sales divided by customers served. Of course this is one way to determine your check average, but it is important to see how this figure compares to the check average you need to meet your daily sales goals. If you are coming in under what you need, you should look at your prices. Check average should be determined by each meal period, especially when different menus are served for each meal. Standards need to be set on how customers who order only a drink and no food are counted.

Seat turnover is how many times you can fill a chair during a meal period with another customer. Bakeries with low check averages need a higher seat turnover to make dine-in sales profitable.

Sales Analysis

Ratio of food to beverage sales is simply the ratio of their percentages of your total sales. In bakeries with a higher percentage of beverage than food sales, profits are generally higher, because there is a greater profit margin on beverages.

Sales mix is the number of each menu item sold. This is crucial to cost analysis because each item impacts food cost differently. If your Wendy's does a huge breakfast business, and the one down the street does a big lunch, your food costs are going to be different than theirs.

Break-even point (BEP) is simply when sales equal expenses, period. Businesses can operate forever at break-even if there are no investors looking for a return on their money.

Profit Analysis

Contribution margin is sales revenue less variable costs.

Closing point is when the cost of being open for a given time period is more expensive than revenue earned. This means that if it cost you $2,000 to open today, and you only made $1,800, your closing point expense will be $200.

Controlling Food Costs

In order to control food costs effectively, there are four things you need to do:

1. Forecast how much and what you are going to sell.
2. Purchase and prepare according to these forecasts.
3. Portion effectively.
4. Control waste and theft.

In order to do these effectively, you must have standards to which you rigorously adhere. Here are two main standards that will help you sustain quality, consistency and low cost:

Standardized Recipes. Since the recipe is the basis for determining the cost of a menu item, standard recipes will assure consistent quality and cost. Standardized recipes include ingredients, preparation methods, yield, used equipment and plate presentation.

Standardized Purchase Specifications are detailed descriptions of the ingredients used in your standardized recipes. Quality and price of all ingredients are known and agreed upon before purchases are made, making the recipe's cost consistent from unit to unit and week to week.

Yield Costs

Once you have standardized recipes in place, you can determine the per plate cost of every dish or individual baked goods. In order to do this, you need to know what the basic ingredients cost and the edible yield of those ingredients for each dish. There are a number of necessary terms for this process:

- **As-Purchased (AP) Weight.** The weight of the product as delivered, including bones, trim, etc.

- **Edible Portion (EP) Weight.** The amount of weight or volume that is available to be portioned after carving or cooking.

- **Waste.** The amount of usable product that is lost due to processing, cooking, or portioning, as well as usable by-products that have no salable value.

- **Usable Trim.** Processing by-products that can be sold as other menu items. These recover a portion or all of their cost.

- **Yield.** The net weight or volume of food after processing but before portioning.

- **Standard Yield.** The yield generated by standardized recipes and portioning procedures—how much usable product there is after processing and cooking.

- **Standard Portion.** The size of the portion according to the standardized recipe, also the basis for determining the cost of the plated portion.

- **Convenience Foods.** Items where at least a portion of the preparation labor is done before delivery. These can include pre-cut chicken, ready-made dough, etc.

These factors allow you to calculate plate costs. The food cost of convenience foods is higher than if you made them from scratch, but once you factor in labor, necessary equipment, inventories of ingredients, more complicated purchasing and storage, etc., you may find that these foods offer considerable savings.

To cost convenience foods you simply count, weigh or measure the portion size and determine how many portions are there. Then divide the number of servable portions into the as-purchased price. Even with their pre-preparation, a small allowance for normal waste must be factored in, often as little as 2 percent per yield.

Costing items from scratch is a little more complex. Most menu items require processing that causes shrinkage of some kind. As a result, if the weight or volume of the cooked product is less than the as-purchased (AP) weight, the edible portion (EP) cost will be higher than the AP price. It's a simple addition of the labor involved and the amount of saleable product being reduced. Through this process, your buyer uses yields to determine quantities to purchase, and your chef discovers optimum quantities to order that result in the highest yield and the least waste.

Sales Mix

Your baked goods menu is where you begin to design a bakery. If you have a specific product "theme" (wedding cakes vs. bagels), your bakery's location must be carefully planned to ensure customer traffic will support your concept. This also works the other way: if you already have the location, design your bread and dessert offerings around the customers you want to attract.

Once your concept and product line are decided, your equipment and kitchen space requirements should be designed around the recipes. Once a kitchen has been built, there is of course some flexibility as you add and remove seasonal items or the latest pastry trend. However, adding new pieces of major equipment may be impossible without high costs or renovations. To design a bakery for today and tomorrow, you need to visualize delivery, processing, preparation, decorating, packaging, and cleaning up. To do this, you must be intimately familiar with each product.

When shopping for equipment, select on the best equipment *for your needs*, not price. Only when you have decided if you need a small fryer or an industrial one, two proofers or five, should you begin to find the best brand and price. This is true for equipment all the way down to pots, pans, dishes and utensils.

Beyond A Loaf of Bread

Your bakery offerings can go beyond a loaf of bread or a chocolate cake, many bakeries have created new profit centers – ready-to-eat breakfast, snack and lunch items that feature their baked goods. A freshly toasted bagel with cream cheese and fresh squeezed, a warm Danish and a cup of coffee or a quick sandwich can increase your sales, bring new customers in and enhance your community recognition.

Start by assessing your ability to handle made-to-order food items – production space, serving utensils and equipment needed and labor costs. Will the espresso machine fit behind the counter? Will making sandwiches make "bread only" customers wait too long?

Selling your baked goods – bread, cookies and pastries – is your core business. Don't over commit your resources to sell prepared snacks and meals. You might start by selling the beverages basics – coffee, tea, soda, milk and an assortment of fruit juices – as they are natural enhancements to your bread and pastry line.

Analyzing your sales mix to determine the impact each item has on sales, costs and profits is an important practice. If you have costs and waste under control, looking at your menu sales mix can help you further reduce costs and boost profits. You will find that some items need to be promoted more aggressively, while others need to be dropped altogether. Classifying your menu items is necessary for making those decisions. Here are some suggested classifications:

1. **Primes.** These are popular items that are low in food cost and high in profit. Have them stand out on your menu.

2. **Standards.** Items with high food costs and high profit margins. You can possibly raise the price on this item and push it as a signature.

3. **Sleepers.** Slow selling low food cost items with low profit margins. Work to increase the likelihood that these will be seen and ordered through more prominent menu display, featuring on menu boards, lowered prices, etc.

4. **Problems.** High in food cost and low in profits. If you can, raise the price and lower production cost. If you can't, hide them on the menu. If sales don't pick up, get rid of them altogether.

Pricing

Pricing is an important aspect of your revenues and customer counts. Prices that are too high will drive customers away, and prices that are too low will kill your profits. However, pricing is not the simple matter of an appropriate markup over cost; it combines other factors as well.

Price can be either *market driven* or *demand driven*. Market driven prices must be responsive to your competitor's prices. Common food items that both you and the place down the road sell need to be priced competitively. This is also true when you're introducing new items where a demand has not been developed. Opposite to these are demand driven items, which customers ask for and where demand exceeds your supply. You have a short-term monopoly on these items, and therefore price is driven up until demand slows or competitors begin to sell similar items.

However you determine your price, the actual marking up of items is an interesting process. A combination of methods is usually a good idea, since each menu item is usually different. Two basic theories are: a) charge as much as you can, and b) charge as little as you can. Each has its pluses and minuses. Obviously, if you charge as much as you can, you increase the chance of greater profits. You do, however, run the risk of needing to offer a product that customers feel is worth the price; otherwise you will lose them because they won't think you're a good value. Charging the lowest price you can gives customers a great sense of value, but lowers your profit margin per item.

Prices are generally determined by competition and demand. Your prices must be in line with the category customers put you in. Burrito joints don't price like a five-star bakery, and vice versa. Both would lose their customer base if they did. While this is an exaggeration, the point is still the same. You want your customers to know your image and your prices to fit into that picture.

Here are four ways to determine prices:

1. **Competitive Pricing.** Simply based on meeting or beating your competition's prices. This is an ineffective method, since it assumes shoppers/diners are making their choice on price alone, and not food quality, ambiance, service, etc.

2. **Intuitive Pricing.** This means you don't want to take the time to find out what your competition is charging, so you are charging based on what you feel guests are willing to pay. If your sense of the value of your product is good, then it works. Otherwise, it can be problematic.

3. **Psychological Pricing.** Price is more of a factor to lower-income customers who go to lower priced bakeries. If they don't know an item is good, they assume it is if it's expensive. If you change your prices, the order in which buyers see them also affects their perceptions. If an item was initially more expensive, it will be viewed as a bargain, and vice versa.

4. **Trial-and-Error Pricing.** This is based on customer reactions to prices. It is not practical in terms of determining your overall prices, but can be effective with individual items to bring them closer to the price a customer is willing to pay, or to distinguish them from similar menu items with a higher or lower food cost.

There are still other factors that help determine prices. Whether customers view you as a leader or a follower can make a big difference on how they view your prices. If people think of you as the best artisan bakery in the area, they'll be willing to pay a little more for a loaf of bread. Service also determines people's sense of value. This is even truer when the difference in actual food quality between you and the competition is negligible. If your customers order at a counter and buss their own tables, this lack of service cost needs to be reflected in your prices. Also, in a competitive market, providing great service can be a factor that puts you in a leadership position and allows you to charge a higher price. Your location, ambience, customer base, product presentation and desired check average all factor into what you feel you can charge and what you need to in order to make a profit.

Financial Analysis

In order to make profits, you need to plan for profits. Many bakeries offering great food, great atmosphere and great service still go out of business. The reason for this is they fail to manage the financial aspects of the business. This means that poor cost control management will be fatal to your business. Furthermore, good financial management is about *interpreting* financial statements and reports, not simply preparing them.

A few distinctions need to be made in order to understand the language we are now using. *Financial accounting* is primarily for external groups to assess taxes, the status of your establishment, etc. *Managerial accounting* provides information to internal users that becomes the basis for managing day-to-day operations. This data is very specific, emphasizes departmental operations and uses non-financial data such as customer counts, menu sales mix and labor hours. These internal reports break down revenues and expenses by department, day and meal period so they can be easily interpreted, and areas that need attention can be seen. Daily and weekly reports must be made and analyzed in order to determine emerging trends.

Shrinkage

What is politely referred to in the retail industry as "shrinkage" is actually theft and fraud. Employee theft accounts for 48.5% of these losses.[2] Other shrinkage categories are shoplifting at 31.3%, administrative error at 15.1% and vendor fraud at 5.1%.

Clearly established and followed controls can reduce this percentage. Begin by separating duties and recording every transaction. If these basic systems are in place, then workers know that they will be held responsible for shrinkage.

In tightly run establishments, cash is more likely to be taken by management than hourly workers, because managers have access to it and know the system well. Hourly workers tend to steal stuff, not cash, because that's what they can get their hands on. Keeping food away from the back door and notifying your employees when you are aware of theft and are investigating can have a deterring effect.

The key to statistical control however, is entering transactions into the system. This can be done electronically or by hand—either way, if food or beverages can be consumed without being entered into the system, your system is flawed, and control is compromised. Five other cost control concepts are crucial to your control system:

1. Documentation of tasks, activities and transactions must be required.

2. Supervision and review of employees by management intimately familiar with set performance standards.

3. Splitting of duties so no single person is involved in all parts of the task cycle.

4. Timeliness. All tasks must be done within set time guidelines; comparisons then made at established control points; and reports made at scheduled times to detect problems.

5. Cost-benefit relationships. Cost of procedures used to benefits gained must exceed the cost of implementing the controls.

The basic control procedure is an independent verification at a control point during and after the completion of a task. This is often done through written or electronic reports. This verification determines if the person performing the task has authority to do so and meets set standards.

---

[2] National Retail Security Survey, November 2002

Point-of-sale systems are also helpful for reducing loss. Once initial training and intimidation are overcome, they can seriously reduce the amount of theft and shrinkage in your bakery.

Purchasing and Ordering

What exactly is the difference? Purchasing is setting the policy on which suppliers, brands, grades and varieties of products will be ordered. These are your standardized purchase specifications; the specifics of how items are delivered, paid for, returned, etc., are negotiated between management and distributors. Basically, purchasing is what you order and from whom. Ordering, then, is simply the act of contacting the suppliers and notifying them of the quantity you require. This is a simpler, lower-level task.

Once menus have been created that meet your customer's satisfaction and your profit needs, a purchasing program designed to assure your profit margins can be developed. An efficient purchasing program incorporates:

- Standard purchase specifications, based on
- Standardized recipes, resulting in
- Standardized yields that, with portion control, allow for
- Accurate costs based on portions actually served.

Buying also has its own distinctions. *Open* or *informal buying* is face-to-face or over-the-phone contact and uses largely oral negotiations and purchase specifics. In *formal buying* terms are put in writing, and payment invoices are stated as conditions for price quotes and customer-service commitments. Its customer service is possibly the most important aspect of the supplier you choose, because good sales representatives know their products, have an understanding of your needs and offer helpful suggestions.

Chapter 23

Controlling Your Labor Costs

L abor costs and turnover are serious concerns in today's bakery market. Increasing labor costs cannot be offset by continuously higher prices without turning customers away. Maximizing worker productivity so few can do more has become a key challenge for food service managers. This is especially important in an industry dominated by an entry-level jobs for the unskilled and uneducated. If qualified applicants are hard to find, you'll need to "create" your own

Manage Time Wisely

Scheduling. The key to controlling labor costs is not a low average hourly wage, but proper scheduling of productive employees. Place your best servers, bakers, etc., where you need them most. This requires knowing the strengths and weaknesses of your employees. Staggering the arrival and departure of employees is a good way to follow the volume of expected customers and minimize labor costs during slow times.

On-Call Scheduling. When your forecasted customer counts are inaccurate, scheduled labor must be adjusted up or down to meet productivity standards. Employees simply wait at home to be called if they are needed for work. If they don't receive a call by a certain time, they know they're not needed. Employees prefer this greatly to coming in only to be sent home when business is slow.

On-Break Schedules. When you can't send someone home, you can put him on a 30-minute break and give him a meal. The 30 minutes is deducted from his timecard, and you can take a credit for the cost of the meal against their minimum wage.

Creating Productivity

A few of the causes of high labor costs and low productivity are poor layout and design of your operation, lack of labor-saving equipment, poor scheduling and no regular detailed system to collect and analyze payroll data. The following are some suggested ways management could improve these areas for greater efficiency:

So often in the food industry, the solution to high labor costs is to lay off employees, lower wages or cut back on hours and/or benefits. These shortsighted measures will initially trim your labor cost, but over time, they will also result in lower overall productivity, a decrease in the quality and service, low morale and a high turnover rate. Occasionally some employees may have to be laid off due to a drastic decrease in sales or to initial over hiring, but this should only rarely occur.

Controlling the bakery's labor cost takes daily management involvement. It cannot be accomplished with one swift action at the end of each month. Described below are some practical suggestions that may be used to streamline your operation so that it may run more efficiently, effectively and profitably.

Design and Equipment

An efficiently designed kitchen with laborsaving equipment is by far the most effective way to reduce labor costs. After several months of operation, examine the kitchen in action. Look at each employee: What are his or her motions and movements? How many steps must be taken to reach food items and more stock? Look at the position and layout of the equipment: Is it set up the most efficient way possible? Ask the employees how they would like their work areas set up, and how work areas could be made more efficient. They are the real experts—they work the same job every day. Look at the service staff's work areas: Could they be made more efficient? These investigations and their results will create faster and better service.

The initial capital expenditures for new equipment can be financed over several years. The cost may be deducted over several years as a tax-deductible depreciation expense. There are also capital investment incentive deductions where you can "write-off" the entire purchase during the first year. Be certain to ask your accountant about these benefits and whether leasing is a better option for you.

Efficient Work Areas

Here are some suggestions on how you might make your kitchen layout work for your production and support staff and your front retail and serving work areas.

- **Break your kitchen activities into self-contained workstations** where ingredients, tools, equipment, supplies and preserving storage are within easy reach.

- **Include plenty of waste receptacles.** Divide by type of waste if you will be implementing recycling programs. Check with your waste management company on local requirements for segregating glass, metal, paper, etc.

- **Create work triangles.** Triangle or diamond layouts give quick access to prep tables, sinks and cooking equipment. Straight-line layouts work best for assembly-line style prep and cooking where more than one person participates.

- **Draw out traffic maps** to minimize unnecessary steps, crisscrossing paths.

- **Allow for ample open space.** People need to pass, carts rolled, shelving moved, large buckets wheeled and trays lifted.

- **Listen to your staff.** Service personnel, bakers and assistants with the hands-on experience can help you create layouts which won't tire them, help them respond quicker and improve morale.

Create Environments That Work With People

Ergonomics is the study and engineering of human physical interaction with spaces and objects during activities. A prep area that requires workers to repeated stretch across to reach ingredients or a frequently used shelf that only very tall workers can safely reach is "poor" ergonomics.

Poor ergonomics = inefficiency = increased labor requirements and increased injury risks = increased costs. Good ergonomics will positively affect your employee's physical well-being, safety, productivity and comfort.

- Eliminate excessive bending, lifting, and reaching while encouraging proper prep and storage procedures.

- Provide stools or chairs to give backs and feet a rest, if the work being done doesn't require standing.

- Create mini-work stations where all necessary food, utensils and prep space are nearby.

- Make certain your tools and equipment weren't designed for only large, tall people. Look for designs that accommodate shorter frames, smaller hands or differing physical characteristics.

- Provide stable, heavy-duty work ladders for accessing top shelves and deep storage units.

- Purchase a supply of important tools and utensils for left handed employees.

- Choose fixtures and equipment that can be easily moved when needed from work area to work area.

Labor-saving Equipment

Every year new pieces of equipment, large and small, expensive and inexpensive, are introduced that will save time, labor and energy. Gone forever are the days when cheap labor will replace the need for new modern equipment. Aside from saving additional labor costs, new mechanization will reduce product handling, eliminate work drudgery and make each task—as well as the overall job—more enjoyable for the employee.

In buying labor-saving equipment, look for equipment that is easy to operate, clean and maintain. If people hate to use it, you may not be saving anything!

Here are a few pieces of equipment that may trim your bakery labor costs:

1. **Combination ovens** – using steam, convection, microwave, cook and hold, dry heat in combination can speed proofing and baking times and eliminate having to transfer goods from equipment to equipment.

2. **Power dishwashers** – minimizes the soak and scrub steps in pot washing.

3. **Food processors** – speed prep time significantly.

4. **Cookie depositor** – transform dough into up to 200 cookies per minute with the Cookie Dough Depositor from Unifiller Systems Inc. For info, visit **http://www.unifiller.com/bakery-machines/cookie-machine**.

5. **Self-service displays and dispensers** – let your customers do the work.

Prepared Foods

Prepared, frozen, portioned and fresh product that is ready to use can save a substantial number of preparation and cooking labor costs. Before committing yourself to using any of these products, inspect a sample to ensure that the product will equal or exceed the desired quality level. Examine the additional food cost of the item.

Could you produce the same product with less overall cost? Remember to consider the additional cost of labor, equipment, utilities and so forth in your projection. Often, the preprepared/manufactured product is considerably cheaper because large quantities are prepared using commercial equipment and procedures. Prepared foods will also contribute to your overall consistency.

There is a big caveat for the independent bakery. Why do people seek you out instead of buying a loaf at the supermarket or picking up a bag of cookies made by those highly-advertised elves or a cheesecake by Sara "what's her name?" *Because they are looking for a fresher and higher quality product.* Your customers probably want artisan breads, elegantly decorated cakes, organic ingredients, real butter and premium chocolate. *Don't lose what makes you unique and worth the drive across town for in trying to save a few dollars.*

Chapter 24

Leaving Your Bakery Business

arely do new business owners consider the importance of being able to sell their business. Because no one has a crystal ball – you cannot predict what life event will cause you to sell your bakery. Build a saleable business is a way to strength your return on your initial and future investments. A profitable bakery can bring top dollar and your profit from the sale is your reward for all the long, hard hours you spent developing it.

The best time to prepare for the future is now! Creating a business that has lasting value means that your hard work will continue to provide you with an excellent in-

come. Your ROI is a healthy retirement nest egg or the capital for a new business venture. Whether you wish to retire, want to move to a new city, or have a life challenge that necessitates you sell your bakery, you'll have better control over the situation if you have a prepared exit plan.

Your Exit Plan

Just as you prepared a written business plan, you should create a short exit plan. Don't forget to review and update your plan annually to reflect your current business state and your objectives.

Your plan should cover:

- **Your exit desires** – best-case scenario. When do you want to retire? Do you want the business to be sold outright or will you let the family continue running it?

- **Your business assessment** – current value. How much could you get in cash if you liquidated or sold it?

- **Ways to enhance your business value.** Have you developed a succession team? Can you increase output or make changes that would make it more attractive to buyers?

- **Preparedness** – worst-case scenario. What will happen in an emergency?

- **Sales preparedness.** Do you know the tax implications of selling? Could you carry the paper?

- **Bowing out.** Do you know what to do to leave your business to others? Dissolve partnerships or corporations?

- **Family preparedness** – securing your family's financial health. Do you have a will? How will they handle your affairs should you die? Will they be able to run the business without you?

Your attorney and accountant can provide valuable advice in creating your plan. For more information on exit plans, visit:

- Principal Financial Group at
  http://www.principal.com/businessowner/bus_exit.htm

- Family Business Experts at
  www.family-business-experts.com/exit-planning.html.

- American Express Small Business at
  www.americanexpress.com/smallbusiness

Passing Your Business On

Millions of small businesses are family-owned and operated. These businesses are passed down through the generations so their legacies continue. Other businesses continue under the guidance of partners or employees.

Issues such as inheritance tax, business trusts, tax-free gifts are all complex issues best left to professionals. Discussing your concerns with your estate planner, banker, accountant and lawyer are critical to ensure a smooth transition and minimize the tax burdens on your family. U.S. Chamber of Commerce offers advice on passing your business on at **www.uschamber.com** as does CCH Business Owners Toolkit has several helpful articles at **http://www.bizfilings.com/toolkit/index.aspx**.

Grooming Your Replacement

Your business exit scenario may mean that someone else steps into your shows. Grooming a replacement takes time – especially if you and your family will continue to have financial ties to your bakery. A natural part of your hiring process should be to envision whether this person might be a good successor for your legacy.

If you have a candidate in mind, start by sharing your vision for the future and develop a plan to:

1. Train on unfamiliar areas.
2. Increase their responsibilities.
3. Review their decision-making abilities.
4. Listen to their needs and ideas.
5. Share money management goals.
6. Set a timeline for transfer.
7. Develop transition stages.
8. Set benchmarks and goals.
9. Examine and improve "problem" issues.
10. Prepare to emotionally and physically leave.

Selling Your Business to Your Employees

You can sell your business to an employee or group of employees as you would with any potential buyer. This is not without risks, as unless they have ample capital, you most likely will be "underwriting" some of the financing.

There are some caveats as feelings may be hurt when your employee/buyers want to make seemingly random changes to "your way of doing things" and your friendships may become strained when haggling over money. Your attorney and accountant can act as a go-between to keep the selling process professional and less emotionally charged.

The Business Law section at **http://www.sba.gov/category/navigation-structure/ starting-managing-business** discusses Employee Owned Stock Plans (EOSP) as an option for transferring your business to your employees. National Center for Employee Ownership offers advice at **www.nceo.org** and the Beyster Institute for Entrepreneurial Employee Ownership at **www.fed.org**.

Transferring your business to a worker co-op can have some advantages for everyone. Worker co-op structures are discussed at the National Cooperative Business Association at http://www.ncba.coop. Transferring your ownership to employees (similar to family inheritance) can also be done.

Saying Good-bye

As an owner/operator of a bakery, saying good-bye to your employees, your early morning schedule and the realization of your dreams can be hard. Letting go is much easier of you have prepared for the day. You've sacrificed and struggled to build a legacy – it is also your responsibility to preserve it!

Chapter 25

Resources

. . . .

Accounting and Financial

- Calculate average costs at **www.cost-watch.com**.
- *Food Cost Control Using Microsoft® Excel(4) for Windows* by Warren Sackler and Samuel R. Trapani.

Food Service and Bakery Associations and Organizations

- American Baker Association—**www.americanbakers.org**
- American Institute of Baking—**www.aibonline.org**
- American Society of Baking—**www.asbe.org**
- Baker's Industry Suppliers Association—**www.bema.org**
- Baking Association of Canada—**www.bakingassoccanada.com**.
- Bread Baker's Guild of America—**www.bbga.org**
- International Cake Exploration Societé—**www. www.ices.org**
- International Dairy- Deli-Bakery Association—**www.iddanet.org**
- National Restaurant Association—**www.restaurant.org**
- Retailer's Bakery Association—**www.rbanet.com**
- San Francisco Baking Institute—**www.sfbi.com**
- Southern Bakers Association—**www.sba.org**

Employee

- Bakery, Confectionery, Tobacco Workers and Grain Millers International Union—**www.bctgm.org**

- Equal Employment Opportunity Commission Web site for employer liability information—**www.eeoc.gov**
- International Union of Bakers—**www.bakeruib.org**

Equipment and Tools

- Auto-Chlor—dishwasher rental and chemical system—**www.autochlorsystem.com**
- Hobart—broad selection of food service equipment—**www.hobartcorp.com**
- Pasty Tools—**www.pastryitems.com**
- Refrigeration information—**www.refrigerators-freezers.com**
- Unifiller.com—filling equipment—**www.unifiller.com**

Music Licensing

- U.S. music licensing agencies: BMI and ASCAP. You can contact ASCAP at **www.ascap.com**, and BMI at **www.bmi.com**. We highly recommend that you contact both BMI and ASCAP to ensure your compliance.

Production

- Baking and Baking Science—**www.bakingandbakingscience.com**

Publications

- BakingBusiness.com—**www.bakingbusiness.com**
- Modern Baking Magazine—**www.bakery-net.com**

Safety, Labeling and Nutritional Info

- Environmental Quality Management Software Program ABA Environmental Compliance Program on Disk—**www.americanbakers.org/pubs/eqmabout.htm**
- Kosher certifications—**http://www.star-k.org/**
- NSF food safety—**http://www.nsf.org/bakery/bakery_cert.html**
- Nutrition Labeling and Education Act—**www.fda.gov/ora/inspect_ref/igs/nleatxt.html**
- The National Organic Program—**www.ams.usda.gov/nop/indexIE.htm**

Index